ST(P) MATHEMATICS 1

ST(P) MATHEMATICS will be completed as follows:

Published 1984	**ST(P) 1**	
	ST(P) 1	Teacher's Notes and Answers
	ST(P) 2	
Published 1985	**ST(P) 2**	Teacher's Notes and Answers
	ST(P) 3A	
	ST(P) 3B	
	ST(P) 3A	Teacher's Notes and Answers
	ST(P) 3B	Teacher's Notes and Answers
Published 1986	**ST(P) 4A**	
	ST(P) 4B	
	ST(P) 4A	Teacher's Notes and Answers
	ST(P) 4B	Teacher's Notes and Answers
Published 1987	**ST(P) 5A**	(with answers)
	ST(P) 5B	(with answers)
Published 1988	**ST(P) 5C**	
	ST(P) 5C	Copy Masters
	ST(P) 5C	Teacher's Notes and Answers
Published 1989	**ST(P)**	Resource Book
Published 1990	**ST(P) 1B**	
	ST(P) 2B	
	ST(P) 1B	Teacher's Notes and Answers
	ST(P) 2B	Teacher's Notes and Answers

ST(P) MATHEMATICS 1

L. Bostock, B.Sc.
Senior Mathematics Lecturer, Southgate Technical College

S. Chandler, B.Sc.
formerly of the Godolphin and Latymer School

A. Shepherd, B.Sc.
Head of Mathematics, Redland High School for Girls

E. Smith, M.Sc.
Head of Mathematics, Tredegar Comprehensive School

Stanley Thornes (Publishers) Ltd

First published in 1984 by:
Stanley Thornes (Publishers) Ltd
Old Station Drive
Leckhampton
CHELTENHAM GL53 0DN
England

Reprinted 1984
Reprinted 1985 with minor corrections
Reprinted 1986
Reprinted 1987
Reprinted 1988
Reprinted 1990

British Library Cataloguing in Publication Data
ST(P) mathematics
  Book 1
  1. Mathematics—1961–
  I. Bostock, L.
  510   QA37.2

ISBN 0 85950 149 3

Typeset by Cotswold Typesetting Ltd, Gloucester
Printed and bound in Great Britain at The Bath Press, Avon

CONTENTS

INTRODUCTION

This book attempts to satisfy your needs as you begin your study of mathematics in the secondary school. We are very conscious of the need for success together with the enjoyment everyone finds in getting things right. With this in mind we have divided most of the exercises into three types of question:

The first type, identified by plain numbers, e.g. **12**, helps you to see if you understand the work. These questions are considered necessary for every chapter you attempt.

The second type, identified by a single underline, e.g. <u>**12**</u>, are extra, but not harder, questions for quicker workers, for extra practice or for later revision.

The third type, identified by a double underline, e.g. <u><u>**12**</u></u>, are for those of you who manage Type 1 questions fairly easily and therefore need to attempt questions that are a little harder.

Most chapters end with "mixed exercises". These will help you revise what you have done, either when you have finished the chapter or at a later date.

All of you should be able to use a calculator accurately by the time you leave school. It is wise, in your first and second years, to use it mainly to check your answers, unless of course you have great difficulty with "tables". Whether you use the calculator or do the working yourself, always estimate your answer and always ask yourself the question, "Is my answer a sensible one?"

1 ADDITION AND SUBTRACTION OF WHOLE NUMBERS

We use whole numbers all the time in everyday life and it is important that we should be able to add them and subtract them accurately in our heads. This comes with practice.

CONTINUOUS ADDITION

To add a line of numbers, start at the left-hand side:

$6+4+3+8 = 21$

Working in your head
add the first two numbers (10)
then add on the next number (13)
then add on the next number (21).
Check your answer by starting at the other end.

To add a column of numbers, start at the bottom and *working in your head* add up the column:

```
  8
  7
  2
+ 5      (5+2 = 7, 7+7 = 14, 14+8 = 22)
───      Check your answer by starting at the top and adding down the
 22      column.
```

EXERCISE 1a Find the value of:

1. $2+3+1+4$
2. $1+5+2+3$
3. $5+2+6+1$
4. $3+4+2+6$
5. $5+6+4+2$

6. $8+2+9+5$
7. $7+3+8+6$
8. $5+4+9+1$
9. $7+3+2+8$
10. $6+7+5+9$

11. $2+5+4+1+3$
12. $4+8+2+1+2$
13. $6+7+3+5+6$
14. $4+9+2+8+4$
15. $7+3+9+6+8$

16. $3+2+3+4+1+5$
17. $4+2+5+6+1+7$
18. $8+3+9+2+7+3$
19. $6+9+4+8+7+5$
20. $4+7+8+6+5+2$

21.		22.		23.		24.		25.	
	3		1		4		9		8
	7		9		6		7		7
	8		5		7		9		6
	+6		+2		+3		+8		+9

26.		27.		28.		29.		30.	
	3		4		6		7		6
	4		2		5		8		7
	5		3		3		2		3
	1		9		1		1		9
	+8		+3		+4		+8		+7

31.		32.		33.		34.		35.	
	3		5		8		2		4
	5		7		7		9		8
	2		3		9		5		2
	9		5		2		8		9
	1		4		8		7		9
	+6		+2		+6		+6		+7

ADDITION OF WHOLE NUMBERS

To add a column of numbers, start with the units:

In the *units* column, $2+1+3 = 6$ so write 6 in the units column.

$$\begin{array}{r} 8\,3 \\ 2\,9\,1 \\ +\,7\,0\,2 \\ \hline 1\,0\,7\,6 \end{array}$$

In the *tens* column, $0+9+8 = 17$ tens which is 7 tens and 1 hundred. Write 7 in the tens column and carry the 1 hundred to the hundreds column to be added to what is there already.

In the *hundreds* column, $1+7+2 = 10$ hundreds which is 0 hundreds and 1 thousand.

EXERCISE 1b   Find the value of the following sums:

1.		2.		3.		4.		5.	
	28		35		22		103		56
	+51		+62		+43		+205		+203

6.		7.		8.		9.		10.	
	101		223		492		259		351
	25		317		812		28		1026
	+273		+342		+735		+704		+ 915

11.		12.		13.		14.		15.	
	87		93		3021		9217		6943
	102		251		84		824		278
	56		179		926		3216		5419
	+304		+1312		+5041		+8572		+3604

$$217+85+976$$

$$217+85+976 = 1278$$

$$\begin{array}{r} 2\,1\,7 \\ 8\,5 \\ +\,9\,7\,6 \\ \hline 1\,2\,7\,8 \\ \hline {\scriptstyle 1\ 1} \end{array}$$

16. $28+72+12$

17. $56+10+92$

18. $83+107+52$

19. $256+139+402$

20. $1026+398+542$

21. $24+83+76$

22. $92+58+27$

23. $52+112+38$

24. $207+394+651$

25. $943+856+984$

26. $826+907+329$

27. $562+497+208$

28. $599+107+2058$

29. $642+321+4973$

30. $555+921+6049$

31. $694+706+293$

32. $325+576+481$

33. $253+431+1212$

34. $821+903+3506$

35. $727+652+2716$

36. $92+56+109+324$

37. $103+72+58+276$

38. $329+26+73+429$

39. $256+82+712+37$

40. $325+293+502+712$

41. $624+1315+437+516$

42. $2514+397+3617+251$

43. $752+593+644+237$

44. $2516+374+527+152$

45. $879+4658+5743+652$

EXERCISE 1c

1. Find the total cost of a tin of baked beans at 17 p, a loaf of bread at 36 p and a can of cola at 16 p.

2. In the local corner shop I bought a comic costing 14 p, a pencil costing 10 p and a packet of sweets costing 15 p. How much did I spend?

3. There are three classes in the first year of a school. One class has 29 children in it, another class has 31 children in it and the third class has 28 children in it. How many children are there in the first year of the school?

4. Find the total cost of a washing machine at £306, a cooker at £257 and a fridge at £194.

5. Write the following numbers in figures:
a) two hundred and sixty-one   b) three hundred and two
c) three thousand and fifty-six   d) thirteen hundred.

6. Write the following numbers in words:
a) 324   b) 5208   c) 150   d) 1500.

7. Add four hundred and fifteen, one hundred and sixty-eight and two hundred and four.

8. I have three pieces of string. One piece is 27 cm long, another piece is 34 cm long and the third piece is 16 cm long. What is the total length of string that I have?

9. Find the total cost of a calculator at £4, a pencil set at £1 and a cassette at £5.

10. When John went to school this morning it took him 4 minutes to walk to the station. He had to wait 12 minutes for the train and the train journey took 26 minutes. He then had an 8 minute walk to his school. How long did it take John to get to school?

11. Find the sum of one thousand and fifty, four hundred and seven and three thousand five hundred.

12. A boy decided to save some money by an unusual method. He put 1 p in his money box the first week, 2 p in the second week, 4 p in the third week, 8 p in the fourth week, and so on. He gave up after 10 weeks. Write down how much he put in his money box each week and add it up to find the total that he had saved. Why do you think he gave up?

SUBTRACTION OF WHOLE NUMBERS

EXERCISE 1d   Do the following subtractions in your head:

| **1.** $\begin{array}{r} 15 \\ -4 \\ \hline \end{array}$ | **2.** $\begin{array}{r} 19 \\ -7 \\ \hline \end{array}$ | **3.** $\begin{array}{r} 18 \\ -4 \\ \hline \end{array}$ | **4.** $\begin{array}{r} 12 \\ -7 \\ \hline \end{array}$ | **5.** $\begin{array}{r} 15 \\ -8 \\ \hline \end{array}$ |

6.	$20-8$	**11.**	$15-2$	**16.**	$11-7$
7.	$18-3$	**12.**	$12-9$	**17.**	$13-8$
8.	$17-8$	**13.**	$17-6$	**18.**	$15-9$
9.	$14-6$	**14.**	$16-8$	**19.**	$20-6$
10.	$10-4$	**15.**	$19-9$	**20.**	$15-7$

You will probably have your own method for subtraction. Use it if you understand it.

Here is one method:

$$\begin{array}{r} \overset{0}{\cancel{8}}\overset{14}{\cancel{0}}\overset{1}{8} \\ -721 \\ \hline 787 \end{array}$$

Start with the units column, then do the tens column and so on. If you cannot do the subtraction, take one from the top number in the next column; this is worth ten in the column on its right.

EXERCISE 1e Find:

$$642-316$$

$$642-316=326$$

$$\begin{array}{r} 6\overset{3}{\cancel{4}}\overset{1}{2} \\ -316 \\ \hline 326 \end{array}$$

$$907-259$$

$$907-259=648$$

$$\begin{array}{r} \overset{8}{\cancel{9}}\overset{9}{\cancel{0}}\overset{1}{7} \\ -259 \\ \hline 648 \end{array}$$

1.	$526-315$	**5.**	$564-491$	**9.**	$283-157$
2.	$754-203$	**6.**	$395-254$	**10.**	$638-452$
3.	$821-415$	**7.**	$708-302$	**11.**	$814-344$
4.	$526-308$	**8.**	$495-369$	**12.**	$592-238$

13.	$578-291$	**17.**	$1237-524$	**21.**	$507-499$
14.	$635-457$	**18.**	$823-568$	**22.**	$3451-623$
15.	$602-415$	**19.**	$718-439$	**23.**	$5267-444$
16.	$704-568$	**20.**	$308-159$	**24.**	$7374-759$

25. $1027 - 452$	29. $4627 - 3924$	33. $3506 - 3429$
26. $3927 - 583$	30. $1203 - 527$	34. $7016 - 6824$
27. $1922 - 398$	31. $4906 - 829$	35. $9342 - 5147$
28. $2704 - 2515$	32. $1516 - 468$	36. $6309 - 4665$

EXERCISE 1f

1. The milk bill for last week was 97 p. I paid with a £5 note (£5 is 500 pence). How much change should I have?

2. In a school there are 856 children. There are 392 girls. How many boys are there?

3. Find the difference between 378 and 293.

4. Take two hundred and fifty-one away from three hundred and forty.

5. A shop starts with 750 cans of cola and sells 463. How many cans are left?

6. Subtract two thousand and sixty-five from eight thousand, five hundred and forty-eight.

7. Find the difference between 182 and 395.

8. The road I live in has 97 houses in it. The road my friend lives in has 49 houses in it. How many more houses are there in my road than in my friend's road?

9. Ben Nevis is 1343 m high and is the highest mountain in Great Britain. Mount Everest is 8843 m high. How much higher than Ben Nevis is Mount Everest?

10. My brother is 123 cm tall and I am 142 cm tall. What is the difference between our heights?

EXERCISE 1g   Find the missing digit; it is marked with □:

1. $27 + 38 = \square 5$
2. $34 + 5\square = 89$
3. $5\square - 25 = 32$
4. $6\square - 48 = 16$
5. $128 + \square 59 = 1087$
6. $5\square + 29 = 83$
7. $\square 4 + 57 = 81$
8. $\square 3 - 47 = 26$
9. $25 - 1\square = 6$
10. $1\square 7 + 239 = 416$

MIXED ADDITION AND SUBTRACTION

It is the sign *in front* of a number that tells you what to do with that number. For example $128-56+92$ means "128 take away 56 and add on 92". This can be done in any order so we could add on 92 and then take away 56, i.e.

$$128-56+92 = 220-56$$
$$= 164$$

EXERCISE 1h Find:

$$138+76-94$$

$$138+76-94 = 214-94$$
$$= 120$$

```
 138
+ 76
 214

 214
- 94
 120
```

$$56-72+39-14$$

$$56-72+39-14 = 95-72-14$$
$$= 23-14$$
$$= 9$$

```
 56
+39
 95

 95
-72
 23
```

1. $25-6+7-9$

2. $14+2-8-3$

3. $7-4+5-6$

4. $19+2-4+3$

5. $23-2+4+5$

6. $46-12+3-9$

7. $27+6-11-9$

8. $2+13-7+3-8$

9. $7-6+9-1-3$

10. $17+4-9-3-5$

11. $17-9+11-19$

12. $36-24+62-49$

13. $51-27-38+14$

14. $43-29+37+16$

15. $124+51-78-14$

16. $91-50+36-27$

17. $105+23-78-50$

18. $73-42-19+27$

19. $215-181+36-70$

20. $361-200+15-81$

21. $213 - 307 + 198 - 31$
22. $29 + 108 - 210 + 93$
23. $493 - 1000 + 751 - 140$
24. $36 + 52 - 73 + 29 - 37$
25. $78 - 43 + 15 - 39 + 18$

26. $612 - 318 + 219 + 84$
27. $95 - 161 + 75 + 10$
28. $952 - 1010 - 251 + 438$
29. $278 + 394 - 506 + 84$
30. $107 - 1127 + 854 + 231$

EXERCISE 1i

1. A boy buys a comic costing 12 p and a pencil costing 8 p. He pays with a 50 p piece. How much change does he get?

2. Find the sum of eighty-six and fifty-four and then take away sixty-eight.

3. I have a piece of string 200 cm long. I cut off two pieces, one of length 86 cm and one of length 34 cm. How long is the piece of string that I have left?

4. On Monday 1000 fish fingers were cooked in the school kitchen. At the first dinner sitting 384 fish fingers were served. At the second sitting 298 fish fingers were served. How many were left?

5. Find the difference between one hundred and ninety and eighty-three. Then add on thirty-seven.

6. A greengrocer has 38 lb of carrots when he opens on Monday morning. During the day he gets a delivery of 60 lb of carrots and sells 29 lb of carrots. How many pounds of carrots are left when he closes on Monday evening?

7. A boy has 30 marbles in his pocket when he goes to school on Monday morning. At first playtime he wins 6 marbles. At second playtime he loses 15 marbles. At third playtime he loses 4 marbles. How many marbles does he now have?

8. What is three hundred and twenty-seven plus two hundred and six minus four hundred and eighty-eight?

9. Sarah gets 50 p pocket money on Saturday. On Monday she spends 34 p. On Tuesday she is given 20 p for doing a special job at home. On Thursday she spends 27 p. How much money has she got left?

10. Make up a problem of your own that involves adding and subtracting numbers.

APPROXIMATION

Display

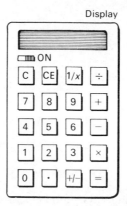

Calculators are very useful and can save a lot of time. Calculators do not make mistakes but *we* sometimes do when we use them. So it is important to know roughly if the answer we get from a calculator is right. By simplifying the numbers involved we can get a rough answer in our heads.

One way to simplify numbers is to make them into the nearest number of tens. For example

127 is roughly 13 tens, or 130

and

123 is roughly 12 tens, or 120

We say that 127 is *rounded up* to 130 and 123 is *rounded down* to 120. In mathematics we say that 127 is approximately equal to 13 tens.

We use the symbol ≈ to mean "is approximately equal to". We would write

127 ≈ 13 tens

123 ≈ 12 tens

When a number is half way between tens we always round up. We say

125 ≈ 13 tens

EXERCISE 1j   Write each of the following numbers as an approximate number of tens:

56 ≈ 6 tens

1. 84	**3.** 46	**5.** 8	**7.** 228	**9.** 73
2. 151	**4.** 632	**6.** 37	**8.** 155	**10.** 4

Write each of the following numbers as an approximate number of hundreds:

$$1278 \approx 13 \text{ hundreds}$$

11.	830	**13.**	780	**15.**	1350	**17.**	1560	**19.**	972
12.	256	**14.**	1221	**16.**	450	**18.**	3780	**20.**	1965

By writing each number correct to the nearest number of tens find an approximate answer for:

$$196 + 58 - 84$$

$$196 \approx 20 \text{ tens}$$
$$58 \approx 6 \text{ tens}$$
$$84 \approx 8 \text{ tens}$$

Therefore $196 + 58 - 84 \approx 20 \text{ tens} + 6 \text{ tens} - 8 \text{ tens}$

$$\approx 18 \text{ tens} = 180$$

21. $344 - 87$ **26.** $89 - 51$

22. $95 - 39$ **27.** $258 + 108$

23. $258 - 49$ **28.** $391 - 127$

24. $472 + 35$ **29.** $275 - 99$

25. $153 + 181$ **30.** $832 - 55$

31. $83 + 27 - 52$ **36.** $49 - 25 + 18$

32. $76 - 31 - 29$ **37.** $68 + 143 + 73$

33. $137 - 56 + 82$ **38.** $153 + 19 + 57$

34. $241 + 37 - 124$ **39.** $369 - 92 + 85$

35. $295 + 304 - 451$ **40.** $250 + 31 - 121$

41. 127+56+82+95   **46.** 83+64+95+51

42. 73+21+37+46+29   **47.** 63+29+40+37+81

43. 33+18+27+96+53   **48.** 108+16+29+53+85

44. 13+29+83+121+5   **49.** 17+23+46+9+75

45. 41+82+96+73+36   **50.** 103+125+76+41+8

Now use your calculator to find the exact answers to numbers 21 to 50. Remember to look at your rough answer to check that your calculator answer is probably right.

2 MULTIPLICATION AND DIVISION OF WHOLE NUMBERS

MULTIPLICATION OF WHOLE NUMBERS

You need to know the multiplication facts, that is, the product of any pair of numbers from the set $\{1, 2, 3, 4, 5, 6, 7, 8, 9\}$. The following exercise will help you to practise the multiplication facts.

For example 69×4 can be found by adding

$$69 + 69 + 69 + 69$$

but it is quicker to use the multiplication facts.

Now $\qquad 69 \times 4 = 9$ units $\times 4 + 6$ tens $\times 4$

9×4 units $= 36$ units $= 3$ tens $+ 6$ units

$$\begin{array}{r} 6\,9 \\ \times \quad 4 \\ \hline 2\,7\,6 \\ \hline {\scriptstyle 3} \end{array}$$

so write down 6 units and carry 3 tens.

6×4 tens $= 24$ tens

Then add on the 3 tens carried to give 27 tens which is 2 hundreds $+ 7$ tens.

EXERCISE 2a Find:

> 24×8
>
> $24 \times 8 = 192$
>
> $$\begin{array}{r} 2\,4 \\ \times \quad 8 \\ \hline 1\,9\,2 \\ \hline {\scriptstyle 3} \end{array}$$

1. 23×2	**4.** 76×4	**7.** 25×4	**10.** 83×5
2. 42×3	**5.** 58×5	**8.** 16×9	**11.** 47×3
3. 13×8	**6.** 31×3	**9.** 72×2	**12.** 54×6
13. 21×6	**16.** 73×4	**19.** 67×8	**22.** 8×21
14. 84×7	**17.** 2×81	**20.** 73×9	**23.** 7×32
15. 36×9	**18.** 33×4	**21.** 49×6	**24.** 9×27

25. 152×4	**28.** 194×2	**31.** 953×3	**34.** 312×7
26. 307×8	**29.** 221×9	**32.** 204×8	**35.** 142×6
27. 256×3	**30.** 211×4	**33.** 876×3	**36.** 513×5
37. 6×529	**40.** 579×9	**43.** 848×8	**46.** 694×8
38. 857×6	**41.** 658×7	**44.** 9×659	**47.** 236×7
39. 7×498	**42.** 7×427	**45.** 748×7	**48.** 573×9

MULTIPLICATION BY 10, 100, 1000, . . .

When 85 is multiplied by 10 the 5 units become 5 tens and the 8 tens become 8 hundreds. So

$$85 \times 10 = 850$$

When 85 is multiplied by 100 the 5 units become 5 hundreds and the 8 tens become 8 thousands. Thus

$$85 \times 100 = 8500$$

When 85 is multiplied by 20 this is the same as $85 \times 2 \times 10$. So

$$85 \times 20 = 85 \times 2 \times 10$$
$$= 170 \times 10$$
$$= 1700$$

In the same way

$$27 \times 4000 = 27 \times 4 \times 1000$$
$$= 108 \times 1000$$
$$= 108\,000$$

EXERCISE 2b

Find 42×900

$$42 \times 900 = 42 \times 9 \times 100$$
$$= 378 \times 100$$
$$= 37\,800$$

$$\begin{array}{r} 42 \\ \times\ \ 9 \\ \hline 378 \\ \scriptstyle 1 \end{array}$$

Find:

1.	27×10	**6.**	27×20	**11.**	73×400
2.	82×100	**7.**	82×300	**12.**	58×60
3.	36×10	**8.**	51×40	**13.**	221×30
4.	108×10	**9.**	39×200	**14.**	127×700
5.	256×1000	**10.**	56×50	**15.**	73×2000

16.	39×900	**21.**	609×80	**26.**	107×400
17.	157×60	**22.**	270×200	**27.**	240×80
18.	295×80	**23.**	556×70	**28.**	100×88
19.	88×70	**24.**	81×3000	**29.**	200×95
20.	350×200	**25.**	390×90	**30.**	856×70

LONG MULTIPLICATION

To multiply 84×26 we use the fact that

$$84 \times 26 = 84 \times 20 + 84 \times 6$$

This can be set out as

$$
\begin{array}{r}
84 \\
\times 26 \\
\hline
504 \\
+1680 \\
\hline
2184 \\
\end{array}
$$

  (84 × 6)
  (84 × 20)

EXERCISE 2c

Find 2813×402

$2813 \times 402 = 1\,130\,826$

$$
\begin{array}{r}
2813 \\
\times \quad 402 \\
\hline
5626 \\
+\,1125200 \\
\hline
1130826 \\
\end{array}
$$

  (2813 × 2)
  (2813 × 400)

Find:

1. 32×21		**6.** 38×41		**11.** 241×32
2. 43×13		**7.** 107×26		**12.** 153×262
3. 86×15		**8.** 53×82		**13.** 433×921
4. 27×21		**9.** 74×106		**14.** 1251×28
5. 34×42		**10.** 36×89		**15.** 3421×33
16. 512×210		**21.** 2004×43		**26.** 385×95
17. 487×82		**22.** 584×97		**27.** 750×450
18. 724×98		**23.** 187×906		**28.** 605×750
19. 146×259		**24.** 270×709		**29.** 1008×908
20. 805×703		**25.** 3060×470		**30.** 1500×802

USING A CALCULATOR FOR LONG MULTIPLICATION

Calculators save a lot of time when used for long multiplication. You do, however, need to be able to estimate the size of answer you expect as a check on your use of the calculator.

One way to get a rough answer is to round off

a number between 10 and 100 to the nearest number of tens
a number between 100 and 1000 to the nearest number of hundreds
a number between 1000 and 10 000 to the nearest number of thousands

and so on.

For example $\quad 512 \times 78 \approx 500 \times 80 = 40\,000$

and $\quad 2752 \times 185 \approx 3000 \times 200 = 600\,000$

EXERCISE 2d Estimate:

1. 79×34	**6.** 59×18	**11.** 159×93
2. 29×27	**7.** 23×55	**12.** 82×309
3. 84×36	**8.** 62×57	**13.** 281×158
4. 45×32	**9.** 136×29	**14.** 631×479
5. 87×124	**10.** 52×281	**15.** 273×784

Estimate the answer and then use your calculator to work out the following:

$$2581 \times 39$$

$$2581 \times 39 \approx 3000 \times 40 = 120\,000 \quad \text{(estimate)}$$
$$2581 \times 39 = 100\,659 \quad \text{(calculator)}$$

16.	258×947	**21.**	78×91	**26.**	52×821
17.	29×384	**22.**	625×14	**27.**	89×483
18.	182×56	**23.**	33×982	**28.**	481×97
19.	37×925	**24.**	2501×12	**29.**	608×953
20.	782×24	**25.**	87×76	**30.**	4897×61

31.	69×78	**36.**	463×87	**41.**	37×634
32.	47×853	**37.**	271×82	**42.**	541×428
33.	94×552	**38.**	753×749	**43.**	798×583
34.	18×47	**39.**	492×47	**44.**	694×7281
35.	62×98	**40.**	68×529	**45.**	7215×48

EXERCISE 2e

1. Multiply three hundred and fifty-six by twenty-three.

2. One jar of marmalade weighs 454 grams. Find the weight of 24 jars.

3. On a school outing 8 coaches were used, each taking 34 children. How many children went on the school outing?

4. A school hall has 30 rows of seats. Each row has 28 seats. How many seats are there?

5. Find the value of one hundred and fifty multiplied by itself.

6. A car park has 34 rows and each row has 42 parking spaces. How many cars can be parked?

7. A supermarket takes delivery of 54 crates of soft drink cans. Each crate contains 48 cans. How many cans are delivered?

8. A school day is 7 hours long. How many minutes are there in the school day?

9. A block of flats has 44 storeys. Each storey has 18 flats. How many flats are there in the block?

10. A light bulb was tested by being left on non-stop. It failed after 28 days exactly. For how many hours was it working?

DIVISION OF WHOLE NUMBERS

$36 \div 8$ means "how many eights are there in 36?". We can find out by repeatedly taking 8 away from 36:

$36 - 8 = 28$

$28 - 8 = 20$

$20 - 8 = 12$ So there are 4 eights in 36 with 4 left over.

$12 - 8 = 4$

Thus $36 \div 8 = 4$, remainder 4.

A quicker way uses the multiplication facts.

We know that $32 = 4 \times 8$

therefore $36 \div 8 = 4$, remainder 4

To find $534 \div 3$ start with the hundreds:

5 (hundreds) $\div$ 3 = $\underline{1}$ (hundred),   remainder 2 (hundreds)

Take the remainder, 2 (hundreds), with the tens:

23 (tens) $\div$ 3 = $\underline{7}$ (tens),   remainder 2 (tens)

Take the remainder, 2 (tens), with the units:

24 (units) $\div$ 3 = $\underline{8}$ units

Therefore $534 \div 3 = 178$

This can be set out as:

$$3 \overline{)\, 5^2 3^2 4} \quad = 178$$

EXERCISE 2f Calculate the following and give the remainder when there is one:

$$4509 \div 5$$

$$\begin{array}{r} 9\ 0\ 1 \quad r\ 4 \\ 5\overline{)\ 4\ 5^0 0^0 9} \end{array}$$

$$4509 \div 5 = 901, \quad r\ 4$$

1.	$87 \div 3$	**6.**	$97 \div 2$	**11.**	$78 \div 8$
2.	$56 \div 4$	**7.**	$73 \div 5$	**12.**	$85 \div 7$
3.	$36 \div 6$	**8.**	$83 \div 4$	**13.**	$39 \div 3$
4.	$57 \div 3$	**9.**	$69 \div 3$	**14.**	$21 \div 9$
5.	$72 \div 4$	**10.**	$82 \div 6$	**15.**	$78 \div 6$

16.	$54 \div 2$	**21.**	$855 \div 5$	**26.**	$294 \div 9$
17.	$639 \div 3$	**22.**	$693 \div 3$	**27.**	$570 \div 7$
18.	$548 \div 2$	**23.**	$721 \div 7$	**28.**	$680 \div 8$
19.	$605 \div 3$	**24.**	$358 \div 5$	**29.**	$731 \div 6$
20.	$497 \div 4$	**25.**	$192 \div 8$	**30.**	$702 \div 5$

31.	$3501 \div 3$	**36.**	$6405 \div 6$	**41.**	$1788 \div 9$
32.	$1763 \div 4$	**37.**	$7399 \div 5$	**42.**	$1098 \div 6$
33.	$4829 \div 2$	**38.**	$8772 \div 4$	**43.**	$2481 \div 7$
34.	$1758 \div 5$	**39.**	$9712 \div 8$	**44.**	$6910 \div 4$
35.	$3852 \div 9$	**40.**	$2009 \div 7$	**45.**	$7505 \div 5$

DIVISION BY 10, 100, 1000, . . .

$812 \div 10$ means "how many tens are there in 812".

There are 81 tens in 810 so

$$812 \div 10 = 81, \quad \text{remainder } 2$$

$2578 \div 100$ means "how many hundreds are there in 2578".

There are 25 hundreds in 2500 so

$$2578 \div 100 = 25, \quad \text{remainder } 78$$

EXERCISE 2g   Calculate the following and give the remainder:

1. $256 \div 10$	**5.** $4910 \div 1000$	**9.** $9426 \div 1000$
2. $87 \div 10$	**6.** $57 \div 10$	**10.** $8512 \div 100$
3. $196 \div 100$	**7.** $186 \div 10$	**11.** $3077 \div 100$
4. $2783 \div 100$	**8.** $2781 \div 10$	**12.** $5704 \div 1000$

LONG DIVISION

To find $2678 \div 21$ we can set the working out as follows:

```
      127
21 ) 2678
     21
     57
     42
    158
    147
     11
```

There is <u>1</u> twenty-one in 26,   r 5 (hundreds).

There are <u>2</u> twenty-ones in 57,   r 15 (tens).

There are <u>7</u> twenty-ones in 158,   r 11 (units).

So $2678 \div 21 = 127$,   r 11.

EXERCISE 2h   Calculate the following and give the remainder:

$2606 \div 25$

$2606 \div 25 = 104$,   r 6

```
      104
25 ) 2606
     25
    106
    100
      6
```

If you use your calculator to check your answers, it will give the whole number part of the answer but it will not give the remainder as a whole number.

1. $254 \div 20$	**5.** $394 \div 19$	**9.** $389 \div 23$
2. $685 \div 13$	**6.** $267 \div 32$	**10.** $298 \div 14$
3. $739 \div 41$	**7.** $875 \div 25$	**11.** $433 \div 15$
4. $862 \div 25$	**8.** $269 \div 16$	**12.** $614 \div 27$

13. $2804 \div 13$	**17.** $2943 \div 23$	**21.** $7514 \div 34$
14. $7315 \div 21$	**18.** $2694 \div 31$	**22.** $5829 \div 43$
15. $8392 \div 34$	**19.** $1875 \div 25$	**23.** $6372 \div 27$
16. $6841 \div 15$	**20.** $3621 \div 30$	**24.** $8261 \div 38$
25. $7315 \div 24$	**29.** $7092 \div 35$	**33.** $5009 \div 60$
26. $8602 \div 15$	**30.** $2694 \div 30$	**34.** $6312 \div 43$
27. $3004 \div 31$	**31.** $8013 \div 40$	**35.** $4321 \div 56$
28. $1608 \div 25$	**32.** $2094 \div 32$	**36.** $7974 \div 17$
37. $103 \div 35$	**41.** $700 \div 28$	**45.** $600 \div 54$
38. $2050 \div 19$	**42.** $4001 \div 36$	**46.** $350 \div 17$
39. $5008 \div 45$	**43.** $3900 \div 43$	**47.** $724 \div 36$
40. $6100 \div 32$	**44.** $2800 \div 14$	**48.** $2390 \div 56$
49. $829 \div 106$	**53.** $3606 \div 300$	**57.** $3742 \div 600$
50. $5241 \div 201$	**54.** $8491 \div 150$	**58.** $8924 \div 120$
51. $3689 \div 151$	**55.** $7625 \div 302$	**59.** $6643 \div 242$
52. $8200 \div 250$	**56.** $8110 \div 400$	**60.** $9260 \div 414$

MIXED OPERATIONS OF $+$, $-$, $\times$, $\div$

When a calculation involves a mixture of the operations $+$, $-$, $\times$, $\div$ we always do

multiplication and division first

For example:
$$2 \times 4 + 3 \times 6 = 8 + 18 \quad \text{(multiplication first)}$$
$$= 26$$

EXERCISE 2i Find:

$$2+3\times6-8\div2$$

$$2+3\times6-8\div2 = 2+18-4 \qquad (\times \text{ and } \div \text{ done first})$$
$$= 16$$

$$5-10\times2\div5+3$$

$$5-10\times2\div5+3 = 5-20\div5+3 \qquad (\times \text{ done})$$
$$= 5-4+3 \qquad (\div \text{ done})$$
$$= 4$$

1. $2+4\times6-8$

2. $24\div8-3$

3. $6+3\times2$

4. $7\times2+6-1$

5. $18\div3-3\times2$

6. $7+4-3\times2$

7. $8\div2+6\times3$

8. $14\times2\div7-3+6$

9. $6-2\times3+7$

10. $5+4\times3+8\div2$

11. $7+3\times2-8\div2$

12. $5-4\div2+7\times2$

13. $6\times3-8\times2$

14. $9\div3+12\div6$

15. $12\div3-15\div5$

16. $9+3-6\div2+1$

17. $6-3\times2+9\div3$

18. $7+2\times4-8\div4$

19. $7\times2+8\times3-2\times6$

20. $5\times3\times2-2\times3\times4$

21. $10\times3\div15+6$

22. $8+7\times4\div2$

23. $3\times8\div4+7$

24. $9\div3+7\times2$

25. $4-8\div2+6$

26. $5\times4\div10+6$

27. $6\times3\div9+2\times4$

28. $7+3\times2\div6$

29. $8\div4+6\div2$

30. $12\div4+3\times2$

31. $19 + 3 \times 2 - 8 \div 2$ **36.** $5 \times 3 \times 4 \div 12 + 6 - 2$

32. $7 \times 2 - 3 + 6 \div 2$ **37.** $5 + 6 \times 2 - 8 \div 2 + 9 \div 3$

33. $8 + 3 \times 2 - 4 \div 2$ **38.** $7 - 9 \div 3 + 6 \times 2 - 4 \div 2$

34. $7 \times 2 - 4 \div 2 + 1$ **39.** $9 \div 3 - 2 + 1 + 6 \times 2$

35. $6 + 8 \div 4 + 2 \times 3 \times 4$ **40.** $4 \times 2 - 6 \div 3 + 3 \times 2 \times 4$

USING BRACKETS

If we need to do some addition and/or subtraction before multiplication and division we use brackets round the section that is to be done first. For example $2 \times (3 + 2)$ means work out $3 + 2$ first.

So $2 \times (3 + 2) = 2 \times 5$

$$= 10$$

For a calculation with brackets and a mixture of $\times$, $\div$, $+$ and $-$ we first work out the inside of the Brackets, then we do the Multiplication and Division, and lastly the Addition and Subtraction.

The capital letters in the last sentence are the same as those in the following sentence:

Bless My Dear Aunt Sally

This should help you remember the order of working.

EXERCISE 2j   Find:

$2 \times (3 \times 6 - 4) + 7 - 12 \div 6$

$2 \times (3 \times 6 - 4) + 7 - 12 \div 6 = 2 \times 14 + 7 - 12 \div 6$ (inside bracket first)

$= 28 + 7 - 2$ ($\times$ and $\div$ next)

$= 33$ (lastly $+$ and $-$)

1. $12 \div (5 + 1)$ **6.** $(3 - 2) \times (5 + 3)$

2. $8 \times (3 + 4)$ **7.** $7 \times (12 - 5)$

3. $(5 - 2) \times 3$ **8.** $(6 + 2) \div 4$

4. $(6 + 1) \times 2$ **9.** $(8 + 1) \times (2 + 3)$

5. $(3 + 2) \times (4 - 1)$ **10.** $(9 - 1) \div (6 - 2)$

11. $2+3\times(3+2)$

12. $7-2\times(5-3)$

13. $8-5+2\times(4+3)$

14. $2\times(7-2)\div(16-11)$

15. $4+3\times(2-1)+8\div(9-7)$

16. $6\div(10-8)+4$

17. $7\times(12-6)-12$

18. $12-8-3\times(9-8)$

19. $4\times(15-7)\div(17-9)$

20. $5\times(8-2)+3\times(7-5)$

21. $6\times8-18\div(2+4)$

22. $10\div5+20\div(4+1)$

23. $5+(2\times10-5)-6$

24. $8-(15\div3+4)+1$

25. $(2\times3-4)+(33\div11+5)$

26. $(18\div3+3)\div(4\times4-7)$

27. $(50\div5+6)-(8\times2-4)$

28. $(10\times3-20)+3\times(9\div3+2)$

29. $(7-3\times2)\div(8\div4-1)$

30. $(5+3)\times2+10\div(8-3)$

EXERCISE 2k

> In the greengrocer's I bought 3 oranges that cost 12 p each and one cabbage that cost 36 p. I paid with a £1 coin. How much change did I get?
>
> $$\begin{aligned}\text{Cost of the oranges} &= 36\,\text{p}\\\text{Cost of the cabbage} &= 36\,\text{p}\\\text{Total cost} &= 72\,\text{p}\\\text{Change from £1} &= 100\,\text{p}-72\,\text{p}\\&= 28\,\text{p}\end{aligned}$$

1. How many apples costing 8 p each can I buy with 50 p?

2. I bought 5 oranges that cost 10 p each and 2 lemons that cost 9 p each. How much did I spend?

3. If a coach holds 30 children how many coaches are needed to take 420 children on a school outing?

4. Three children went into a sweet shop. The first child bought three sweets costing 2 p each, the second child bought three sweets costing 1 p each and the third child bought three sweets costing 3 p each. How much money did they spend altogether?

5. A girl saves the same amount each week. After 8 weeks she has 64 p. How much does she save each week?

6. I bought five stamps at 17 p each. How much change did I get from 100 p?

7. A car travelling at 50 miles an hour took 3 hours to travel from London to Cardiff. How many miles did the car travel?

8. A club started the year with 82 members. During the year 36 people left and 28 people joined. How many people belonged to the club at the end of the year?

9. One money box has five 5 p pieces and four 10 p pieces in it. Another money box has six 10 p pieces and ten 2 p pieces in it. What is the total sum of money in the two money boxes?

10. A greengrocer bought a sack of potatoes weighing 50 kg. He divided the potatoes into bags, so that each bag held 3 kg of potatoes. How many complete bags of potatoes did he get from his sack?

11. At a school election one candidate got 26 votes, and the other candidate got 35 votes. 10 voting papers were spoiled and 5 pupils did not vote. How many children could have voted altogether?

12. I bought five pencils costing 12 p each. How much change did I get from £1?

13. Three children are given 60 p to split equally amongst them. How much does each child get?

14. A person can walk up a flight of steps at the rate of 30 steps a minute. It takes him 3 minutes to reach the top. How many steps are there?

15. An extension ladder is made of three separate parts, each 300 cm long. There is an overlap of 30 cm at each junction when it is fully extended. How long is the extended ladder?

16. Jane, Sarah and Claire come to school with 20 p each. Jane owes Sarah 10 p and she also owes Claire 5 p. Sarah owes Jane 4 p and she also owes Claire 8 p. When all their debts are settled, how much money does each girl have?

17. At the newsagent's I buy two comics costing 14 p each, and a magazine costing 50 p. How much change do I get from £1?

18. A man gets paid £105 for a five day working week. How much does he get paid a day?

19. The total number of children in the first year of a school is 500. There are 50 more girls than boys. How many of each are there?

20. 4000 apples are packed into boxes, each box holding 75 apples. How many boxes are required?

21. In a book of street plans of a town, the street plans start on page 6 and end on page 72. How many pages of street plans are there?

22. My great-grandmother died in 1894, aged 62. In which year was she born?

23. How many times can 5 be taken away from 132?

24. An oak tree was planted in the year in which Lord Swell was born. He died in 1940, aged 80. How old was the oak tree in 1984?

25. A mountaineer starts from a point which is 150 m above sea level. He climbs 200 m and then descends 50 m before climbing another 300 m. How far is he now above sea level?

26. A bus leaves the bus station at 9.30 a.m. It reaches the Town Hall at 9.40 a.m. and gets to the railway station at 9.52 a.m. How long does it take to go from the Town Hall to the railway station?

27. A class is told to work out the odd numbered questions in an exercise containing 30 questions. How many questions do they have to do?

28. In the hardware shop I bought 3 screws that cost 5 p each and 2 light bulbs that cost 30 p each. I paid with a 50 p piece and two 20 p pieces. How much change did I get?

29. A vegetable plot is 1000 cm long. Cabbages are planted in a row down the length of the plot. If the cabbages are planted 30 cm apart and the first cabbage is planted 5 cm from the end, how many cabbages can be planted in one row?

30. A train timetable reads as follows:

Euston	depart	9.30 a.m.
Rugby	arrive	10.30 a.m.
	depart	10.35 a.m.
Birmingham	arrive	11.00 a.m.

How long does the journey from Euston to Rugby take?
How long does the journey from Rugby to Birmingham take?

31. Write down a question that uses the same numbers and gives the same answer as the problem in question 28.

NUMBER PATTERNS

EXERCISE 2I

2	7	6
9	5	1
4	3	8

This is a magic square.
The numbers in every row, in every column and in each diagonal add up to 15.

Copy and complete the following magic squares. Use the numbers 1 to 9 just once in each, and use a pencil in case you need to rub out!

1.

8		
	5	
4		2

2.

4	9	
	5	
	1	

3. Use each of the numbers 1 to 16 just once to complete the 4 × 4 magic square below. Each row, column and diagonal should add up to 34.

		7	
15			
9	5	16	
8		1	13

4. Make up a 3 × 3 magic square of your own. Use the numbers 1 to 9 just once each and put 5 in the middle.

5. 1, 3, 5, 7, . . .   This is a sequence. By looking at it you should be able to find the rule for getting the next number.

In this sequence, the next number is always 2 bigger than the number before it.
Write down the next two numbers in this sequence.

In questions 6 to 17 write down the next two numbers in the sequence:

6. 1, 4, 7, 10,   **10.** 3, 6, 9, 12,   **14.** 1, 10, 100, 1000,

7. 12, 10, 8, 6,   **11.** 64, 32, 16, 8,   **15.** 81, 72, 63, 54,

8. 1, 5, 9, 13,   **12.** 1, 3, 9, 27,   **16.** 3, 7, 11, 15,

9. 2, 4, 8, 16,   **13.** 4, 9, 16, 25,   **17.** 5, 10, 17, 26,

18. Consider the following pattern:

$$1 \qquad\qquad = 1 = 1 \times 1$$
$$1+3 \qquad\quad = 4 = 2 \times 2$$
$$1+3+5 \quad\; = 9 = 3 \times 3$$
$$1+3+5+7 = 16 = 4 \times 4$$

Write down the next three lines in this pattern.
Now try and write down (without adding them up) the sum of
a) the first eight odd numbers   b) the first twenty odd numbers.

19. Consider the following pattern:

$$2 \qquad\qquad = 2 = 1 \times 2$$
$$2+4 \qquad\quad = 6 = 2 \times 3$$
$$2+4+6 \quad\; = 12 = 3 \times 4$$
$$2+4+6+8 = 20 = 4 \times 5$$

Write down the next three lines in this pattern.
How many consecutive even numbers, beginning with 2, have a sum
of 156? ($156 = 12 \times 13$).

20.

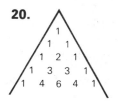

Try to find the pattern in the given triangle of
numbers. Can you write down the next three
rows?

21.

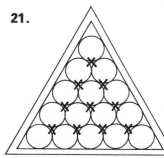

Fifteen red snooker balls are placed in the
frame as shown. A second layer is then placed
on top so that they rest on these in the spaces
marked with crosses. This is followed by more
layers until there is a single ball at the top of
the pyramid. How many balls are needed to
make this pyramid?

22. This is a class game. Start with a number and then each pupil in turn
adds on a fixed number to the last number called. For example if you
start with 5 and each pupil adds on 4 to the last number called, it will
go 5, 9, 13, 17, ... If you make a mistake you are out.

23. A different version of the game in question 22 is to start with a fairly high number and then each pupil in turn subtracts a fixed number from the last one called.

MIXED EXERCISES

EXERCISE 2m Find:

1. $126 + 501 + 378$ **4.** $84 \div 3$ **7.** $35 + 86 + 94 + 27$

2. $153 - 136$ **5.** $350 + 8796 - 2538$ **8.** $20 \div (9 - 4) + 3$

3. 76×9 **6.** $8 \times 321 - 1550$

9. How many packets of popcorn costing 15 p each can I buy with £1?

10. I buy three bars of chocolate costing 18 p each. How much change do I get from £1?

EXERCISE 2n Find:

1. $92 + 625 + 153$ **4.** $79 \div 8$ **7.** $68 - 42 + 12 \times 2$

2. $247 - 193$ **5.** $(7 + 30) \times 2 - 45$ **8.** $79 - 35 + 56 - 63$

3. 84×8 **6.** $382 - 792 \div 3$

9. How many times can 6 be taken away from 45?

10. The contents of a tin of sweets weighs 2500 grams. The sweets are divided into packets each weighing 500 grams. How many packets of sweets can be made up?

EXERCISE 2p Find:

1. $296 + 1025 + 983$ **4.** 106×32 **7.** $940 + 360 - 1040$

2. $347 - 84$ **5.** $2501 \div 9$ **8.** $2983 \div 150$

3. 7×59 **6.** $7863 \div 20$

9. A youth club has 80 members. There are 10 more boys than girls. How many of each are there?

10. There were two candidates in a school election and they got 25 votes and 32 votes. 10 voting papers were spoiled. If 100 children could have voted, how many children did not vote?

EXERCISE 2q   Find:

1. $749 + 821 + 1563$ **4.** 284×16 **7.** $15 - 4 \times (12 - 9)$

2. $278 - 109$ **5.** $2781 \div 10$ **8.** $54 + (7 \times 8 - 10) + 32$

3. 205×40 **6.** $728 - 180 \div 12$

9. I buy three stamps costing 21 p each. How much change do I get from £1?

10. Write down a question which uses the same numbers and gives the same answer as the problem in question 9.

3 FRACTIONS: ADDITION AND SUBTRACTION

THE MEANING OF FRACTIONS

Think of cutting a cake right through the middle into two equal pieces. Each piece is one half of the cake. One half is a fraction, written as $\frac{1}{2}$.

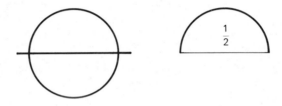

If we cut the cake into four equal pieces, each piece is one quarter, written $\frac{1}{4}$, of the cake. When one piece is taken away there are three pieces left, so the fraction that is left is three quarters, or $\frac{3}{4}$.

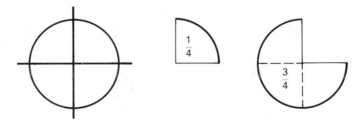

When the cake is divided into five equal slices, one slice is $\frac{1}{5}$, two slices is $\frac{2}{5}$, three slices is $\frac{3}{5}$ and four slices is $\frac{4}{5}$ of the cake.

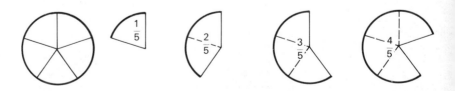

Notice that the top number in each fraction (called the *numerator*) tells you *how many* slices and the bottom number (called the *denominator*) tells you about the size of the slices.

30

EXERCISE 3a In each of the following sketches, write down the fraction that is shaded:

1.

4.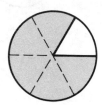

2.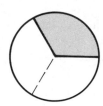

5.

3.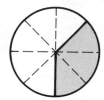

6.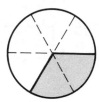

It is not only cakes that can be divided into fractions. Anything at all that can be split up can be divided into fractions.

Write down the fraction that is shaded in each of the following diagrams:

7.

9.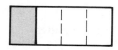

8.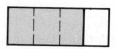

10.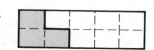

11.

14.

12.

15.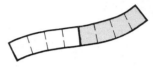

13.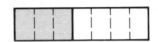

16.

ONE QUANTITY AS A FRACTION OF ANOTHER

Quite a lot of things are divided into equal parts. For instance a week is divided into seven days, so each day is $\frac{1}{7}$ of a week. One pound is divided into one hundred pence, so each penny is $\frac{1}{100}$ of a pound.

EXERCISE 3b

> In June there were 23 sunny days. What fraction of June was sunny?
>
> There are 30 days in June.
>
> $$23 \text{ sunny days} = \frac{23}{30} \text{ of June}$$

1. One hour is divided into 60 minutes. What fraction of an hour is

   a) one minute   b) nine minutes   c) thirty minutes
   d) forty-five minutes?

2. You go to school on five days each week. What fraction of a week is this?

3. In the month of January, snow fell on eleven days. What fraction of all the days in January had snow falls?

In questions 4 to 13 write the first quantity as a fraction of the second quantity:

> Write 10 minutes as a fraction of 1 hour.
>
> (We must always use the same unit for both quantities. This time we will use minutes, so we want to write 10 minutes as a fraction of 60 minutes.)
>
> $$10 \text{ minutes} = \tfrac{10}{60} \text{ of 1 hour}$$

4. 51 days; 1 year (not a leap year)

5. 35 p; £1

6. 90 p; £5

7. 35 seconds; 3 minutes

8. 3 days; the month of January

9. 17 days; the months of June and July together

10. 5 days; 3 weeks

11. £1.50; £5

12. 45 minutes; 2 hours

13. 37 seconds; 1 hour

14. A boy gets 80 p pocket money. If he spends 45 p, what fraction of his pocket money is left?

15. In a class of thirty-two children, ten take French, eight take music and twenty-five take geography. What fraction of the children in the class take

a) French   b) music   c) geography?

16. A girl's journey to school costs 15 p on one bus and 25 p on another bus. What fraction of the total cost arises from each bus?

17. In an orchard there are twenty apple trees, eighteen plum trees, fourteen cherry trees and ten pear trees. What fraction of all the trees are

a) apple trees   b) pear trees   c) *not* cherry trees?

18. In a Youth Club with 37 members, 12 are more than 15 years old and 8 are under 14 years old. What fraction of the members are

a) over 15   b) under 14   c) 14 and over?

19. During a summer holiday of fourteen days there were three rainy days, two cloudy days and all the other days were sunny. What fraction of the holiday was

a) sunny   b) rainy?

EQUIVALENT FRACTIONS

In the first sketch below, a cake is cut into four equal pieces. One slice is $\frac{1}{4}$ of the cake.

In the second sketch the cake is cut into eight equal pieces. Two slices is $\frac{2}{8}$ of the cake.

In the third sketch the cake is cut into sixteen equal slices. Four slices is $\frac{4}{16}$ of the cake.

But the same amount of cake has been taken each time.

Therefore
$$\frac{1}{4} = \frac{2}{8} = \frac{4}{16}$$

and we say that $\frac{1}{4}$, $\frac{2}{8}$ and $\frac{4}{16}$ are *equivalent fractions*.

Now $\dfrac{1}{4} = \dfrac{1 \times 2}{4 \times 2} = \dfrac{2}{8}$ and $\dfrac{1}{4} = \dfrac{1 \times 4}{4 \times 4} = \dfrac{4}{16}$

So all we have to do to find equivalent fractions is to multiply the numerator and the denominator by the same number. For instance

$$\frac{1}{4} = \frac{1 \times 3}{4 \times 3} = \frac{3}{12}$$

and $$\frac{1}{4} = \frac{1 \times 5}{4 \times 5} = \frac{5}{20}$$

Any fraction can be treated in this way.

EXERCISE 3c   In questions 1 to 6 draw cake diagrams to show that:

1. $\dfrac{1}{3} = \dfrac{2}{6}$ **3.** $\dfrac{1}{5} = \dfrac{2}{10}$ **5.** $\dfrac{2}{3} = \dfrac{6}{9}$

2. $\dfrac{1}{2} = \dfrac{3}{6}$ **4.** $\dfrac{3}{4} = \dfrac{9}{12}$ **6.** $\dfrac{2}{3} = \dfrac{8}{12}$

In questions 7 to 33 fill in the missing numbers to make equivalent fractions:

$$\frac{1}{5} = \frac{3}{-}$$

(If $\frac{1}{5} = \frac{3}{-}$ the numerator has been multiplied by 3)

$$\frac{1}{5} = \frac{1 \times 3}{5 \times 3} = \frac{3}{15}$$

$$\frac{1}{5} = \frac{}{20}$$

(If $\frac{1}{5} = \frac{}{20}$ the denominator has been multiplied by 4)

$$\frac{1}{5} = \frac{1 \times 4}{5 \times 4} = \frac{4}{20}$$

---

7. $\dfrac{1}{3} = \dfrac{2}{-}$ **11.** $\dfrac{1}{6} = \dfrac{3}{-}$ **15.** $\dfrac{9}{10} = \dfrac{90}{-}$

8. $\dfrac{2}{5} = \dfrac{}{10}$ **12.** $\dfrac{1}{3} = \dfrac{}{12}$ **16.** $\dfrac{1}{6} = \dfrac{}{36}$

9. $\dfrac{3}{7} = \dfrac{9}{-}$ **13.** $\dfrac{2}{5} = \dfrac{6}{-}$ **17.** $\dfrac{4}{5} = \dfrac{}{20}$

10. $\dfrac{9}{10} = \dfrac{}{40}$ **14.** $\dfrac{3}{7} = \dfrac{}{28}$ **18.** $\dfrac{2}{3} = \dfrac{12}{-}$

19. $\dfrac{2}{9} = \dfrac{4}{-}$

20. $\dfrac{3}{8} = \dfrac{}{80}$

21. $\dfrac{5}{11} = \dfrac{}{22}$

22. $\dfrac{4}{5} = \dfrac{8}{-}$

23. $\dfrac{1}{10} = \dfrac{10}{-}$

24. $\dfrac{2}{9} = \dfrac{}{36}$

25. $\dfrac{3}{8} = \dfrac{}{800}$

26. $\dfrac{5}{11} = \dfrac{50}{-}$

27. $\dfrac{4}{5} = \dfrac{}{50}$

28. $\dfrac{1}{10} = \dfrac{100}{-}$

29. $\dfrac{2}{9} = \dfrac{20}{-}$

30. $\dfrac{3}{8} = \dfrac{3000}{-}$

31. $\dfrac{5}{11} = \dfrac{}{121}$

32. $\dfrac{4}{5} = \dfrac{400}{-}$

33. $\dfrac{1}{10} = \dfrac{1000}{-}$

34. Write each of the following fractions as an equivalent fraction with denominator 24:

> $\dfrac{2}{3}$
>
> $$\dfrac{2}{3} = \dfrac{2 \times 8}{3 \times 8} = \dfrac{16}{24}$$

a) $\dfrac{1}{2}$ b) $\dfrac{1}{3}$ c) $\dfrac{1}{6}$ d) $\dfrac{3}{4}$ e) $\dfrac{5}{12}$ f) $\dfrac{3}{8}$

35. Write each of the following fractions in equivalent form with denominator 45:

a) $\dfrac{2}{15}$ b) $\dfrac{4}{9}$ c) $\dfrac{3}{5}$ d) $\dfrac{1}{3}$ e) $\dfrac{14}{15}$ f) $\dfrac{1}{5}$

36. Find an equivalent fraction with denominator 36 for each of the following fractions:

a) $\dfrac{3}{4}$ b) $\dfrac{5}{9}$ c) $\dfrac{1}{6}$ d) $\dfrac{5}{18}$ e) $\dfrac{7}{12}$ f) $\dfrac{2}{3}$

37. Change each of the following fractions into an equivalent fraction with numerator 12:

a) $\dfrac{1}{6}$ b) $\dfrac{3}{4}$ c) $\dfrac{6}{7}$ d) $\dfrac{4}{5}$ e) $\dfrac{2}{3}$ f) $\dfrac{1}{2}$

38. Some of the following equivalent fractions are correct but two of them are wrong. Find the wrong ones and correct them by altering the numerator:

a) $\dfrac{2}{5} = \dfrac{6}{15}$ b) $\dfrac{2}{3} = \dfrac{4}{9}$ c) $\dfrac{3}{7} = \dfrac{6}{14}$

d) $\dfrac{4}{9} = \dfrac{12}{27}$ e) $\dfrac{7}{10} = \dfrac{77}{100}$

COMPARING THE SIZES OF FRACTIONS

Suppose that we want to see which is bigger, $\frac{5}{7}$ or $\frac{2}{3}$. Before we can compare these two fractions we must change them into the *same kind* of fraction. That means we must find equivalent fractions that have the same denominator. This denominator must be a number that both 7 and 3 divide into. So our new denominator is 21. Now

$$\frac{5}{7} = \frac{15}{21} \qquad \text{and} \qquad \frac{2}{3} = \frac{14}{21}$$

We can see that $\frac{15}{21}$ is bigger than $\frac{14}{21}$, i.e. $\frac{5}{7}$ is bigger than $\frac{2}{3}$.

We often use the symbol $>$ instead of writing the words "is bigger than". Using this symbol we could write $\frac{15}{21} > \frac{14}{21}$, so $\frac{5}{7} > \frac{2}{3}$.

Similarly we use $<$ instead of writing "is less than".

EXERCISE 3d In the following questions find which is the bigger fraction:

$\dfrac{3}{5}$ or $\dfrac{7}{11}$

$\dfrac{3}{5} = \dfrac{33}{55}$ and $\dfrac{7}{11} = \dfrac{35}{55}$ (55 divides by 5 and by 11)

So $\frac{7}{11}$ is the bigger fraction.

1. $\dfrac{1}{2}$ or $\dfrac{1}{3}$ **3.** $\dfrac{2}{3}$ or $\dfrac{4}{5}$ **5.** $\dfrac{2}{7}$ or $\dfrac{3}{8}$

2. $\dfrac{3}{4}$ or $\dfrac{5}{6}$ **4.** $\dfrac{2}{9}$ or $\dfrac{1}{7}$ **6.** $\dfrac{2}{3}$ or $\dfrac{3}{4}$

7. $\dfrac{2}{5}$ or $\dfrac{3}{7}$ **9.** $\dfrac{3}{8}$ or $\dfrac{1}{5}$ **11.** $\dfrac{3}{5}$ or $\dfrac{4}{7}$

8. $\dfrac{5}{6}$ or $\dfrac{3}{5}$ **10.** $\dfrac{4}{5}$ or $\dfrac{6}{7}$ **12.** $\dfrac{3}{4}$ or $\dfrac{2}{3}$

13. $\dfrac{1}{4}$ or $\dfrac{3}{11}$ **17.** $\dfrac{1}{4}$ or $\dfrac{2}{7}$ **21.** $\dfrac{9}{11}$ or $\dfrac{7}{9}$

14. $\dfrac{5}{7}$ or $\dfrac{3}{5}$ **18.** $\dfrac{5}{8}$ or $\dfrac{4}{7}$ **22.** $\dfrac{2}{5}$ or $\dfrac{1}{3}$

15. $\dfrac{3}{8}$ or $\dfrac{5}{11}$ **19.** $\dfrac{2}{9}$ or $\dfrac{3}{11}$ **23.** $\dfrac{4}{7}$ or $\dfrac{3}{5}$

16. $\dfrac{3}{10}$ or $\dfrac{4}{11}$ **20.** $\dfrac{5}{7}$ or $\dfrac{7}{9}$ **24.** $\dfrac{5}{8}$ or $\dfrac{6}{11}$

In questions 25 to 36, put either $>$ or $<$ between the fractions:

$$\frac{2}{5} \qquad \frac{3}{7}$$

$$\frac{2}{5} = \frac{14}{35} \quad \text{and} \quad \frac{3}{7} = \frac{15}{35} \qquad \text{(35 divides by 5 and by 7)}$$

So $\frac{2}{5} < \frac{3}{7}$.

25. $\dfrac{1}{4}$ $\dfrac{2}{7}$ **29.** $\dfrac{3}{10}$ $\dfrac{1}{4}$ **33.** $\dfrac{4}{9}$ $\dfrac{5}{11}$

26. $\dfrac{2}{3}$ $\dfrac{5}{8}$ **30.** $\dfrac{1}{3}$ $\dfrac{2}{5}$ **34.** $\dfrac{2}{11}$ $\dfrac{1}{7}$

27. $\dfrac{3}{7}$ $\dfrac{1}{2}$ **31.** $\dfrac{3}{5}$ $\dfrac{2}{3}$ **35.** $\dfrac{8}{11}$ $\dfrac{3}{4}$

28. $\dfrac{5}{8}$ $\dfrac{7}{10}$ **32.** $\dfrac{2}{9}$ $\dfrac{1}{5}$ **36.** $\dfrac{7}{8}$ $\dfrac{7}{9}$

Arrange the following fractions in ascending order:

$$\frac{3}{4}, \frac{7}{10}, \frac{1}{2}, \frac{4}{5}$$

$$\frac{3}{4} = \frac{15}{20}$$

$$\frac{7}{10} = \frac{14}{20}$$

(20 divides by 4, 10, 2 and 5)

$$\frac{1}{2} = \frac{10}{20}$$

$$\frac{4}{5} = \frac{16}{20}$$

So the ascending order is $\frac{1}{2}$, $\frac{7}{10}$, $\frac{3}{4}$, $\frac{4}{5}$.

37. $\dfrac{2}{3}, \dfrac{1}{2}, \dfrac{3}{5}, \dfrac{7}{30}$

40. $\dfrac{2}{5}, \dfrac{3}{8}, \dfrac{17}{20}, \dfrac{1}{2}, \dfrac{7}{10}$

38. $\dfrac{13}{20}, \dfrac{3}{4}, \dfrac{4}{10}, \dfrac{5}{8}$

41. $\dfrac{5}{7}, \dfrac{11}{14}, \dfrac{3}{4}, \dfrac{17}{28}, \dfrac{1}{2}$

39. $\dfrac{1}{3}, \dfrac{5}{6}, \dfrac{1}{2}, \dfrac{7}{12}$

42. $\dfrac{7}{10}, \dfrac{2}{5}, \dfrac{3}{5}, \dfrac{14}{25}, \dfrac{1}{2}$

Arrange the following fractions in descending order:

43. $\dfrac{5}{6}, \dfrac{1}{2}, \dfrac{7}{9}, \dfrac{11}{18}, \dfrac{2}{3}$

46. $\dfrac{7}{10}, \dfrac{11}{15}, \dfrac{2}{3}, \dfrac{23}{30}, \dfrac{4}{5}$

44. $\dfrac{13}{20}, \dfrac{3}{5}, \dfrac{1}{2}, \dfrac{3}{4}, \dfrac{7}{10}$

47. $\dfrac{7}{16}, \dfrac{1}{2}, \dfrac{5}{8}, \dfrac{19}{32}, \dfrac{3}{4}$

45. $\dfrac{7}{12}, \dfrac{1}{6}, \dfrac{2}{3}, \dfrac{17}{24}, \dfrac{3}{4}$

48. $\dfrac{4}{5}, \dfrac{7}{12}, \dfrac{5}{6}, \dfrac{1}{2}, \dfrac{3}{4}$

SIMPLIFYING FRACTIONS

Think of the way you find equivalent fractions. For example

$$\frac{2}{5} = \frac{2 \times 7}{5 \times 7} = \frac{14}{35}$$

Looking at this the other way round we see that

$$\frac{14}{35} = \frac{\not7 \times 2}{\not7 \times 5} = \frac{2}{5}$$

In the middle step, 7 is a factor of both the numerator and the denominator and it is called a common factor. To get the final value of $\frac{2}{5}$ we have "crossed out" the common factor and this is called cancelling. What we have really done is to divide the top and the bottom by 7 and this *simplifies* the fraction.

When all the simplifying is finished we say that the fraction is in its *lowest terms*.

Any fraction whose numerator and denominator have a common factor (perhaps more than one) can be simplified in this way. Suppose, for example, that we want to simplify $\frac{24}{27}$. As 3 is a factor of 24 and of 27, we say

$$\frac{24}{27} = \frac{3 \times 8}{3 \times 9} = \frac{8}{9}$$

A quicker way to write this down is to divide the numerator and the denominator mentally by the common factor, crossing them out and writing the new numbers beside them (it is a good idea to write the new numbers smaller so that you can see that you have simplified the fraction), i.e.

$$\frac{\not{24}^{\,8}}{\not{27}_{\,9}} = \frac{8}{9}$$

EXERCISE 3e   Simplify the following fractions:

$$\frac{66}{176}$$

$$\frac{\not{66}^{\,33\,3}}{\not{176}_{\,88\,8}} = \frac{3}{8}$$

(We divided top and bottom by 2 and then by 11.)

1. $\frac{2}{6}$ **3.** $\frac{3}{9}$ **5.** $\frac{9}{27}$ **7.** $\frac{5}{15}$ **9.** $\frac{10}{20}$

2. $\frac{30}{50}$ **4.** $\frac{6}{12}$ **6.** $\frac{4}{8}$ **8.** $\frac{12}{18}$ **10.** $\frac{8}{32}$

11. $\dfrac{8}{28}$ **13.** $\dfrac{14}{70}$ **15.** $\dfrac{16}{56}$ **17.** $\dfrac{36}{72}$ **19.** $\dfrac{60}{100}$

12. $\dfrac{27}{90}$ **14.** $\dfrac{24}{60}$ **16.** $\dfrac{10}{30}$ **18.** $\dfrac{15}{75}$ **20.** $\dfrac{36}{90}$

21. $\dfrac{70}{126}$ **23.** $\dfrac{99}{132}$ **25.** $\dfrac{80}{100}$ **27.** $\dfrac{54}{162}$ **29.** $\dfrac{27}{36}$

22. $\dfrac{49}{77}$ **24.** $\dfrac{33}{121}$ **26.** $\dfrac{48}{84}$ **28.** $\dfrac{54}{66}$ **30.** $\dfrac{800}{1000}$

ADDING FRACTIONS

Suppose there is a bowl of oranges and apples. First you take three oranges and then two more oranges. You then have five oranges; we can add the 3 and the 2 together because they are the same kind of fruit. But three oranges and two apples cannot be added together because they are different kinds of fruit.

For fractions it is the denominator that tells us the kind of fraction, so we can add fractions together if they have the same denominator but not while their denominators are different.

EXERCISE 3f   Add the fractions, simplifying the answers where you can:

$\dfrac{2}{7} + \dfrac{3}{7}$

$$\dfrac{2}{7} + \dfrac{3}{7} = \dfrac{2+3}{7}$$

$$= \dfrac{5}{7}$$

$\dfrac{9}{22} + \dfrac{5}{22}$

$$\dfrac{9}{22} + \dfrac{5}{22} = \dfrac{9+5}{22}$$

$$= \dfrac{\cancel{14}^{7}}{\cancel{22}_{11}}$$

$$= \dfrac{7}{11}$$

Add the fractions given in questions 1 to 24, simplifying the answers where you can:

1. $\dfrac{1}{4}+\dfrac{2}{4}$

5. $\dfrac{11}{23}+\dfrac{8}{23}$

9. $\dfrac{2}{21}+\dfrac{9}{21}$

2. $\dfrac{1}{8}+\dfrac{3}{8}$

6. $\dfrac{1}{7}+\dfrac{2}{7}$

10. $\dfrac{7}{30}+\dfrac{8}{30}$

3. $\dfrac{3}{11}+\dfrac{2}{11}$

7. $\dfrac{2}{5}+\dfrac{1}{5}$

11. $\dfrac{6}{13}+\dfrac{5}{13}$

4. $\dfrac{3}{13}+\dfrac{7}{13}$

8. $\dfrac{3}{10}+\dfrac{1}{10}$

12. $\dfrac{1}{10}+\dfrac{7}{10}$

13. $\dfrac{2}{7}+\dfrac{4}{7}$

17. $\dfrac{5}{16}+\dfrac{7}{16}$

21. $\dfrac{4}{11}+\dfrac{2}{11}$

14. $\dfrac{4}{17}+\dfrac{5}{17}$

18. $\dfrac{8}{19}+\dfrac{3}{19}$

22. $\dfrac{14}{23}+\dfrac{1}{23}$

15. $\dfrac{3}{14}+\dfrac{4}{14}$

19. $\dfrac{3}{20}+\dfrac{7}{20}$

23. $\dfrac{11}{18}+\dfrac{5}{18}$

16. $\dfrac{8}{30}+\dfrac{19}{30}$

20. $\dfrac{21}{100}+\dfrac{19}{100}$

24. $\dfrac{7}{15}+\dfrac{3}{15}$

We can add more than two fractions in the same way.

Add the fractions given in questions 25 to 34:

$$\dfrac{3}{17}+\dfrac{5}{17}+\dfrac{8}{17}$$

$$\dfrac{3}{17}+\dfrac{5}{17}+\dfrac{8}{17}=\dfrac{3+5+8}{17}$$

$$=\dfrac{16}{17}$$

25. $\dfrac{2}{15}+\dfrac{4}{15}+\dfrac{6}{15}$

28. $\dfrac{1}{14}+\dfrac{3}{14}+\dfrac{5}{14}+\dfrac{2}{14}$

26. $\dfrac{8}{100}+\dfrac{21}{100}+\dfrac{11}{100}$

29. $\dfrac{2}{51}+\dfrac{4}{51}+\dfrac{6}{51}+\dfrac{8}{51}+\dfrac{7}{51}$

27. $\dfrac{3}{31}+\dfrac{2}{31}+\dfrac{7}{31}+\dfrac{11}{31}$

30. $\dfrac{3}{19}+\dfrac{2}{19}+\dfrac{7}{19}$

31. $\dfrac{7}{60}+\dfrac{8}{60}+\dfrac{11}{60}$

33. $\dfrac{3}{100}+\dfrac{14}{100}+\dfrac{31}{100}+\dfrac{2}{100}$

32. $\dfrac{4}{45}+\dfrac{11}{45}+\dfrac{8}{45}+\dfrac{2}{45}$

34. $\dfrac{3}{99}+\dfrac{11}{99}+\dfrac{4}{99}+\dfrac{7}{99}$

FRACTIONS WITH DIFFERENT DENOMINATORS

To add fractions with different denominators we must first change the fractions into equivalent fractions with the same denominator. This new denominator must be a number that both original denominators divide into. For instance, if we want to add $\frac{2}{5}$ and $\frac{3}{7}$ we choose 35 for our new denominator because 35 can be divided by both 5 and 7:

$$\frac{2}{5}=\frac{14}{35}$$

$$\frac{3}{7}=\frac{15}{35}$$

So
$$\frac{2}{5}+\frac{3}{7}=\frac{14}{35}+\frac{15}{35}=\frac{29}{35}$$

EXERCISE 3g Find:

> $$\frac{2}{7}+\frac{3}{8}$$
>
> (7 and 8 both divide into 56)
>
> $$\frac{2}{7}+\frac{3}{8}=\frac{16}{56}+\frac{21}{56}=\frac{37}{56}$$

1. $\dfrac{2}{3}+\dfrac{1}{5}$ **5.** $\dfrac{3}{10}+\dfrac{2}{3}$ **9.** $\dfrac{1}{6}+\dfrac{2}{7}$

2. $\dfrac{1}{5}+\dfrac{3}{8}$ **6.** $\dfrac{4}{7}+\dfrac{1}{8}$ **10.** $\dfrac{5}{6}+\dfrac{1}{7}$

3. $\dfrac{1}{5}+\dfrac{1}{6}$ **7.** $\dfrac{3}{7}+\dfrac{1}{6}$ **11.** $\dfrac{3}{11}+\dfrac{5}{9}$

4. $\dfrac{2}{5}+\dfrac{3}{7}$ **8.** $\dfrac{2}{3}+\dfrac{2}{7}$ **12.** $\dfrac{2}{9}+\dfrac{3}{10}$

The new denominator, which is called the *common denominator*, is not always as big as you might first think. For instance, if we want to add $\frac{3}{4}$ and $\frac{1}{12}$, the common denominator is 12 because it divides by both 4 and 12.

$$\frac{3}{4}+\frac{1}{12}$$

$$\frac{3}{4}+\frac{1}{12}=\frac{9}{12}+\frac{1}{12}$$

$$=\frac{10^5}{12^6}$$

$$=\frac{5}{6}$$

---

13. $\frac{2}{5}+\frac{3}{10}$

14. $\frac{3}{8}+\frac{7}{16}$

15. $\frac{3}{7}+\frac{8}{21}$

16. $\frac{3}{10}+\frac{3}{100}$

17. $\frac{1}{4}+\frac{7}{10}$

18. $\frac{1}{4}+\frac{3}{8}$

19. $\frac{2}{3}+\frac{2}{9}$

20. $\frac{4}{9}+\frac{5}{18}$

21. $\frac{1}{20}+\frac{3}{5}$

22. $\frac{4}{11}+\frac{5}{22}$

23. $\frac{2}{5}+\frac{7}{15}$

24. $\frac{7}{12}+\frac{1}{6}$

More than two fractions can be added in this way. The common denominator must be divisible by *all* of the original denominators.

$$\frac{1}{8}+\frac{1}{2}+\frac{1}{3}$$

(8, 2 and 3 all divide into 24)

$$\frac{1}{8}+\frac{1}{2}+\frac{1}{3}=\frac{3}{24}+\frac{12}{24}+\frac{8}{24}$$

$$=\frac{3+12+8}{24}$$

$$=\frac{23}{24}$$

25. $\dfrac{1}{5}+\dfrac{1}{4}+\dfrac{1}{2}$ **29.** $\dfrac{1}{7}+\dfrac{3}{14}+\dfrac{1}{2}$ **33.** $\dfrac{7}{20}+\dfrac{3}{10}+\dfrac{1}{5}$

26. $\dfrac{1}{8}+\dfrac{1}{4}+\dfrac{1}{3}$ **30.** $\dfrac{1}{3}+\dfrac{1}{6}+\dfrac{1}{2}$ **34.** $\dfrac{2}{9}+\dfrac{2}{3}+\dfrac{1}{18}$

27. $\dfrac{3}{10}+\dfrac{2}{5}+\dfrac{1}{4}$ **31.** $\dfrac{1}{2}+\dfrac{3}{8}+\dfrac{1}{10}$ **35.** $\dfrac{2}{15}+\dfrac{1}{10}+\dfrac{2}{5}$

28. $\dfrac{5}{12}+\dfrac{1}{6}+\dfrac{1}{3}$ **32.** $\dfrac{1}{3}+\dfrac{2}{9}+\dfrac{1}{6}$ **36.** $\dfrac{1}{4}+\dfrac{1}{12}+\dfrac{1}{3}$

SUBTRACTING FRACTIONS

Exactly the same method is used for subtracting fractions as for adding them. To work out the value of $\frac{7}{8}-\frac{3}{8}$ we notice that the denominators are the same, so

$$\frac{7}{8}-\frac{3}{8}=\frac{7-3}{8}$$
$$=\frac{4}{8}$$
$$=\frac{1}{2}$$

EXERCISE 3h

Find $\dfrac{7}{9}-\dfrac{1}{4}$

(The denominators are not the same so we use equivalent fractions with denominator 36.)

$$\frac{7}{9}-\frac{1}{4}=\frac{28}{36}-\frac{9}{36}$$
$$=\frac{28-9}{36}$$
$$=\frac{19}{36}$$

Find:

1. $\dfrac{8}{9} - \dfrac{2}{9}$

5. $\dfrac{9}{10} - \dfrac{1}{2}$

9. $\dfrac{2}{3} - \dfrac{3}{7}$

2. $\dfrac{7}{10} - \dfrac{2}{10}$

6. $\dfrac{5}{7} - \dfrac{2}{7}$

10. $\dfrac{4}{7} - \dfrac{1}{3}$

3. $\dfrac{6}{17} - \dfrac{1}{17}$

7. $\dfrac{8}{13} - \dfrac{3}{13}$

11. $\dfrac{11}{15} - \dfrac{4}{15}$

4. $\dfrac{3}{4} - \dfrac{1}{5}$

8. $\dfrac{19}{20} - \dfrac{7}{20}$

12. $\dfrac{13}{18} - \dfrac{7}{18}$

13. $\dfrac{8}{11} - \dfrac{2}{5}$

17. $\dfrac{19}{100} - \dfrac{1}{10}$

21. $\dfrac{3}{4} - \dfrac{5}{8}$

14. $\dfrac{7}{9} - \dfrac{2}{3}$

18. $\dfrac{5}{8} - \dfrac{2}{7}$

22. $\dfrac{7}{12} - \dfrac{1}{3}$

15. $\dfrac{8}{13} - \dfrac{1}{2}$

19. $\dfrac{15}{16} - \dfrac{3}{4}$

23. $\dfrac{13}{18} - \dfrac{5}{9}$

16. $\dfrac{11}{12} - \dfrac{5}{6}$

20. $\dfrac{7}{15} - \dfrac{1}{5}$

24. $\dfrac{13}{15} - \dfrac{3}{5}$

ADDING AND SUBTRACTING FRACTIONS

Fractions can be added and subtracted in one problem in a similar way. For example

$$\frac{7}{9} + \frac{1}{18} - \frac{1}{6} = \frac{14}{18} + \frac{1}{18} - \frac{3}{18}$$

$$= \frac{14 + 1 - 3}{18}$$

$$= \frac{12}{18}$$

$$= \frac{2}{3}$$

It is not always possible to work from left to right in order because we have to subtract too much too soon. In this case we can do the adding first. Remember that it is the sign *in front* of a number that tells you what to do with that number.

EXERCISE 3i Find:

$$\frac{1}{8} - \frac{3}{4} + \frac{11}{16}$$

$$\frac{1}{8} - \frac{3}{4} + \frac{11}{16} = \frac{2}{16} - \frac{12}{16} + \frac{11}{16}$$

$$= \frac{13 - 12}{16}$$

$$= \frac{1}{16}$$

1. $\dfrac{3}{4} + \dfrac{1}{2} - \dfrac{7}{8}$

5. $\dfrac{3}{5} + \dfrac{3}{25} - \dfrac{27}{50}$

9. $\dfrac{7}{10} - \dfrac{41}{100} + \dfrac{1}{20}$

2. $\dfrac{6}{7} - \dfrac{9}{14} + \dfrac{1}{2}$

6. $\dfrac{2}{3} + \dfrac{1}{6} - \dfrac{5}{12}$

10. $\dfrac{5}{8} - \dfrac{21}{40} + \dfrac{2}{5}$

3. $\dfrac{3}{8} + \dfrac{7}{16} - \dfrac{3}{4}$

7. $\dfrac{4}{5} - \dfrac{7}{10} + \dfrac{1}{2}$

11. $\dfrac{7}{12} - \dfrac{1}{6} + \dfrac{1}{3}$

4. $\dfrac{11}{12} + \dfrac{1}{6} - \dfrac{2}{3}$

8. $\dfrac{7}{9} - \dfrac{2}{3} + \dfrac{5}{6}$

12. $\dfrac{2}{3} - \dfrac{7}{18} + \dfrac{2}{9}$

13. $\dfrac{2}{9} - \dfrac{1}{3} + \dfrac{1}{6}$

17. $\dfrac{1}{6} - \dfrac{5}{18} + \dfrac{1}{3}$

21. $\dfrac{3}{10} - \dfrac{61}{100} + \dfrac{1}{2}$

14. $\dfrac{1}{6} - \dfrac{2}{3} + \dfrac{7}{12}$

18. $\dfrac{1}{5} - \dfrac{7}{10} + \dfrac{17}{20}$

22. $\dfrac{1}{8} - \dfrac{7}{24} + \dfrac{5}{12}$

15. $\dfrac{2}{5} - \dfrac{1}{2} + \dfrac{3}{10}$

19. $\dfrac{1}{4} - \dfrac{5}{8} + \dfrac{1}{2}$

23. $\dfrac{1}{3} - \dfrac{5}{18} + \dfrac{2}{9}$

16. $\dfrac{1}{8} - \dfrac{13}{16} + \dfrac{3}{4}$

20. $\dfrac{2}{3} - \dfrac{5}{6} + \dfrac{1}{2}$

24. $\dfrac{3}{10} + \dfrac{2}{15} - \dfrac{2}{5}$

PROBLEMS

EXERCISE 3j

In a class of school children, $\frac{1}{3}$ of the children come to school by bus, $\frac{1}{4}$ come to school on bicycles and the rest walk to school. What fraction of the children ride to school? What fraction do not use a bus?

The fraction who ride to school on bicycle and bus $= \dfrac{1}{3}+\dfrac{1}{4}$

$$= \dfrac{4+3}{12}$$

$$= \dfrac{7}{12}$$

Therefore $\frac{7}{12}$ of the children ride to school.

The complete class of children is a whole unit, i.e. 1.

The fraction of children who do not use a bus is found by taking the bus users from the complete class, i.e.

$$\dfrac{1}{1}-\dfrac{1}{3}=\dfrac{3-1}{3}=\dfrac{2}{3}$$

1. A girl spends $\frac{1}{5}$ of her pocket money on sweets and $\frac{2}{3}$ on records. What fraction has she spent? What fraction has she left?

2. A group of friends went to a hamburger bar. $\frac{2}{5}$ of them bought a hamburger, $\frac{1}{3}$ of them just bought chips. The rest bought cola. What fraction of the group bought food? What fraction bought a drink?

3. At a pop festival, $\frac{2}{3}$ of the groups were all male, $\frac{1}{4}$ of the groups had one girl and the rest had more than one girl. What fraction of the groups

a) were not all male   b) had more than one girl?

4. At a Youth Club, $\frac{1}{2}$ of the meetings are for playing table tennis, $\frac{1}{8}$ of the meetings are discussions and the rest are record sessions. What fraction of the meetings are

a) record sessions   b) not for discussions?

5. At a school, $\frac{1}{8}$ of the time is spent in mathematics classes, $\frac{3}{20}$ of the time in English classes and $\frac{1}{20}$ on games. What fraction of the time is spent on

a) English and maths together   b) all lessons except games

c) maths and games?

MIXED NUMBERS AND IMPROPER FRACTIONS

Most of the fractions we have met so far have been less than a whole unit. These are called *proper* fractions. But we often have more than a whole unit. Suppose, for instance, that we have one and a half bars of chocolate:

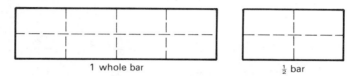

1 whole bar $\frac{1}{2}$ bar

We have $1\frac{1}{2}$ bars, and $1\frac{1}{2}$ is called a *mixed number.*

Another way of describing the amount of chocolate is to say that we have three half bars.

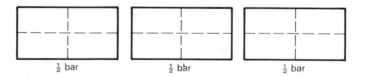

$\frac{1}{2}$ bar $\frac{1}{2}$ bar $\frac{1}{2}$ bar

We have $\frac{3}{2}$ bars and $\frac{3}{2}$ is called an *improper* fraction because the numerator is bigger than the denominator.

But the amount of chocolate in the two examples is the same, so

$$\frac{3}{2} = 1\frac{1}{2}$$

Improper fractions can be changed into mixed numbers by finding out how many whole units there are. For instance, to change $\frac{8}{3}$ into a mixed number we look for the biggest number below 8 that divides by 3, i.e. 6. Then

$$\frac{8}{3} = \frac{6+2}{3} = \frac{6}{3} + \frac{2}{3} = 2 + \frac{2}{3} = 2\frac{2}{3}$$

EXERCISE 3k   In questions 1 to 20 change the improper fractions into mixed numbers:

$$\frac{15}{4}$$

$$\frac{15}{4} = \frac{12+3}{4}$$

$$= \frac{12}{4} + \frac{3}{4}$$

$$= 3\frac{3}{4}$$

1. $\frac{9}{4}$ **3.** $\frac{37}{6}$ **5.** $\frac{88}{9}$ **7.** $\frac{27}{4}$ **9.** $\frac{127}{5}$

2. $\frac{19}{4}$ **4.** $\frac{53}{10}$ **6.** $\frac{7}{2}$ **8.** $\frac{41}{8}$ **10.** $\frac{114}{11}$

11. $\frac{109}{8}$ **13.** $\frac{121}{9}$ **15.** $\frac{87}{11}$ **17.** $\frac{41}{3}$ **19.** $\frac{73}{3}$

12. $\frac{83}{7}$ **14.** $\frac{91}{6}$ **16.** $\frac{77}{6}$ **18.** $\frac{67}{5}$ **20.** $\frac{49}{10}$

We can also change mixed numbers into improper fractions. For instance, in $2\frac{4}{5}$ we have two whole units and $\frac{4}{5}$. In each whole unit there are five fifths, so in $2\frac{4}{5}$ we have ten fifths and four fifths, i.e.

$$2\frac{4}{5} = \frac{10}{5} + \frac{4}{5} = \frac{14}{5}$$

EXERCISE 3l   In questions 1 to 20, change the mixed numbers into improper fractions:

$$3\frac{1}{7}$$

$$3\frac{1}{7} = 3 + \frac{1}{7}$$

$$= \frac{21}{7} + \frac{1}{7}$$

$$= \frac{22}{7}$$

1. $4\frac{1}{3}$ **3.** $1\frac{7}{10}$ **5.** $8\frac{1}{7}$ **7.** $2\frac{6}{7}$ **9.** $3\frac{2}{3}$

2. $8\frac{1}{4}$ **4.** $10\frac{8}{9}$ **6.** $6\frac{3}{5}$ **8.** $4\frac{1}{6}$ **10.** $5\frac{1}{2}$

11. $7\frac{2}{5}$ **13.** $3\frac{4}{5}$ **15.** $8\frac{3}{4}$ **17.** $1\frac{9}{10}$ **19.** $7\frac{3}{8}$

12. $2\frac{4}{9}$ **14.** $4\frac{7}{9}$ **16.** $10\frac{3}{7}$ **18.** $6\frac{2}{3}$ **20.** $10\frac{1}{10}$

THE MEANING OF 15 ÷ 4

$15 \div 4$ means "how many fours are there in 15?".

There are 3 fours in 15 with 3 left over, so $15 \div 4 = 3$, remainder 3.

Now that remainder, 3, is $\frac{3}{4}$ of 4. Thus we can say that there are $3\frac{3}{4}$ fours in 15

i.e. $$15 \div 4 = 3\frac{3}{4}$$

But $$\frac{15}{4} = 3\frac{3}{4}$$

Therefore

$15 \div 4$ and $\frac{15}{4}$ mean the same thing

EXERCISE 3m Calculate the following divisions, giving your answers as mixed numbers:

> $27 \div 8$
>
> $$27 \div 8 = \frac{27}{8}$$
> $$= 3\frac{3}{8}$$

1. $36 \div 7$	**5.** $82 \div 5$	**9.** $98 \div 12$
2. $59 \div 6$	**6.** $29 \div 4$	**10.** $107 \div 10$
3. $52 \div 11$	**7.** $41 \div 3$	**11.** $37 \div 5$
4. $20 \div 8$	**8.** $64 \div 9$	**12.** $52 \div 8$

ADDING MIXED NUMBERS

If we want to find the value of $2\frac{1}{3} + 3\frac{1}{4}$ we add the whole numbers and then the fractions, i.e.

$$2\frac{1}{3} + 3\frac{1}{4} = 2 + 3 + \frac{1}{3} + \frac{1}{4}$$

$$= 5 + \frac{4+3}{12}$$

$$= 5 + \frac{7}{12}$$

$$= 5\frac{7}{12}$$

Sometimes there is an extra step in the calculation. For example

$$3\frac{1}{2} + 2\frac{3}{8} + 5\frac{1}{4} = 3 + 2 + 5 + \frac{1}{2} + \frac{3}{8} + \frac{1}{4}$$

$$= 10 + \frac{4+3+2}{8}$$

$$= 10 + \frac{9}{8}$$

But $\frac{9}{8}$ is an improper fraction, so we change it into a mixed number

i.e.
$$3\frac{1}{2} + 2\frac{3}{8} + 5\frac{1}{4} = 10 + \frac{8+1}{8}$$

$$= 10 + 1 + \frac{1}{8}$$

$$= 11\frac{1}{8}$$

EXERCISE 3n   Find:

1. $2\frac{1}{4} + 3\frac{1}{2}$

2. $1\frac{1}{2} + 2\frac{1}{3}$

3. $4\frac{1}{5} + 1\frac{3}{8}$

4. $5\frac{1}{9} + 4\frac{1}{3}$

5. $3\frac{1}{4} + 2\frac{5}{9}$

6. $1\frac{1}{3} + 2\frac{5}{6}$

7. $3\frac{1}{4} + 1\frac{1}{5}$

8. $2\frac{1}{7} + 1\frac{1}{14}$

9. $6\frac{3}{10} + 1\frac{2}{5}$

10. $8\frac{1}{7} + 5\frac{2}{3}$

11. $7\frac{3}{8} + 3\frac{7}{16}$

12. $1\frac{3}{4} + 4\frac{7}{12}$

16. $2\frac{7}{10} + 9\frac{1}{5}$

17. $5\frac{7}{10} + 2\frac{3}{5}$

13. $3\frac{5}{7} + 7\frac{1}{2}$

18. $9\frac{2}{3} + 8\frac{5}{6}$

14. $6\frac{1}{2} + 1\frac{9}{16}$

19. $2\frac{4}{5} + 7\frac{3}{10}$

15. $8\frac{7}{8} + 3\frac{3}{16}$

20. $6\frac{3}{10} + 4\frac{4}{5}$

21. $1\frac{1}{4} + 3\frac{2}{3} + 6\frac{7}{12}$

26. $4\frac{3}{5} + 8\frac{7}{10} + 2\frac{1}{2}$

22. $5\frac{1}{7} + 4\frac{1}{2} + 7\frac{11}{14}$

27. $3\frac{7}{10} + 9\frac{21}{100} + 1\frac{3}{5}$

23. $3\frac{3}{4} + 5\frac{1}{8} + 8\frac{5}{16}$

28. $4\frac{1}{4} + 7\frac{1}{8} + 6\frac{1}{32}$

24. $10\frac{2}{3} + 3\frac{1}{6} + 7\frac{2}{9}$

29. $1\frac{5}{7} + 11\frac{1}{2} + 9\frac{1}{14}$

25. $4\frac{4}{5} + 9\frac{4}{15} + 1\frac{1}{3}$

30. $10\frac{7}{9} + 6\frac{1}{3} + 5\frac{7}{18}$

SUBTRACTING MIXED NUMBERS

If we want to find the value of $5\frac{3}{4} - 2\frac{2}{5}$ we can use the same method as for adding:

$$5\frac{3}{4} - 2\frac{2}{5} = 5 - 2 + \frac{3}{4} - \frac{2}{5}$$

$$= 3 + \frac{15 - 8}{20}$$

$$= 3 + \frac{7}{20}$$

$$= 3\frac{7}{20}$$

But when we find the value of $6\frac{1}{4} - 2\frac{4}{5}$ we get

$$6\frac{1}{4} - 2\frac{4}{5} = 6 - 2 + \frac{1}{4} - \frac{4}{5}$$

$$= 4 + \frac{1}{4} - \frac{4}{5}$$

This time it is not so easy to deal with the fractions because $\frac{4}{5}$ is bigger than $\frac{1}{4}$. So we take one of the whole units and change it into a fraction, giving

$$3 + 1 + \frac{1}{4} - \frac{4}{5}$$

$$= 3 + \frac{20 + 5 - 16}{20}$$

$$= 3 + \frac{9}{20}$$

$$= 3\frac{9}{20}$$

EXERCISE 3p Find:

1. $2\frac{3}{4} - 1\frac{1}{8}$ **5.** $7\frac{3}{4} - 2\frac{1}{3}$ **9.** $4\frac{4}{5} - 3\frac{1}{10}$

2. $3\frac{2}{3} - 1\frac{4}{5}$ **6.** $3\frac{5}{6} - 2\frac{1}{3}$ **10.** $6\frac{5}{7} - 3\frac{2}{5}$

3. $1\frac{5}{6} - \frac{2}{3}$ **7.** $2\frac{6}{7} - 1\frac{1}{2}$ **11.** $3\frac{1}{3} - 1\frac{1}{5}$

4. $3\frac{1}{4} - 2\frac{1}{2}$ **8.** $4\frac{1}{2} - 2\frac{1}{5}$ **12.** $5\frac{3}{4} - 2\frac{1}{2}$

13. $8\frac{4}{5} - 5\frac{1}{2}$ **17.** $7\frac{1}{2} - 5\frac{3}{4}$ **21.** $8\frac{6}{7} - 5\frac{3}{4}$

14. $5\frac{7}{9} - 3\frac{5}{7}$ **18.** $4\frac{3}{5} - 1\frac{1}{4}$ **22.** $3\frac{1}{2} - 1\frac{7}{8}$

15. $4\frac{5}{8} - 1\frac{1}{3}$ **19.** $7\frac{6}{7} - 4\frac{3}{5}$ **23.** $2\frac{1}{2} - 1\frac{3}{4}$

16. $6\frac{3}{4} - 3\frac{6}{7}$ **20.** $8\frac{8}{11} - 2\frac{2}{3}$ **24.** $5\frac{4}{7} - 3\frac{4}{5}$

25. $3\frac{1}{4} - 1\frac{7}{8}$ **29.** $8\frac{2}{3} - 7\frac{8}{9}$ **33.** $9\frac{7}{10} - 5\frac{4}{5}$

26. $5\frac{3}{5} - 2\frac{9}{10}$ **30.** $4\frac{1}{6} - 2\frac{2}{3}$ **34.** $2\frac{5}{12} - 1\frac{3}{4}$

27. $9\frac{7}{10} - 6\frac{1}{5}$ **31.** $6\frac{2}{3} - 3\frac{5}{6}$ **35.** $4\frac{7}{9} - 3\frac{11}{18}$

28. $6\frac{3}{10} - 3\frac{4}{5}$ **32.** $7\frac{3}{4} - 4\frac{7}{8}$ **36.** $5\frac{1}{3} - 2\frac{4}{7}$

MIXED EXERCISES

EXERCISE 3q **1.** Calculate:

a) $\frac{2}{3} + \frac{4}{7}$ b) $\frac{5}{6} - \frac{3}{8}$ c) $\frac{3}{8} + \frac{1}{9}$ d) $1\frac{1}{2} + \frac{2}{3}$ e) $2\frac{1}{4} - 1\frac{1}{3}$

2. Simplify:

a) $\frac{54}{24}$ b) $3\frac{15}{75}$

3. Write the first quantity as a fraction of the second quantity:

a) 3 days; 1 week   b) 17 children; 30 children

4. Write the following fractions in ascending size order:

a) $\frac{1}{2}, \frac{7}{10}, \frac{3}{5}, \frac{13}{20}$ b) $\frac{3}{4}, \frac{7}{12}, \frac{5}{6}, \frac{2}{3}$ c) $\frac{3}{5}, \frac{7}{10}, \frac{17}{20}, \frac{71}{100}$

5. Write either > or < between the following pairs of fractions:

a) $\frac{5}{12}$ $\frac{7}{16}$ b) $\frac{3}{8}$ $\frac{7}{24}$ c) $\frac{13}{22}$ $\frac{19}{33}$

6. A cricket club consists of 7 good batsmen, 5 good bowlers, 4 all-rounders and some non-players. If there are 22 people in the club, what fraction of them are

a) non-players   b) good batsmen   c) not all-rounders?

EXERCISE 3r **1.** Calculate:

a) $\frac{4}{5} - \frac{2}{3}$

b) $\frac{9}{10} + \frac{4}{5}$

c) $\frac{7}{11} - \frac{1}{2}$

d) $2\frac{1}{3} + 4\frac{1}{4}$

e) $3\frac{1}{6} - 2\frac{2}{3}$

f) $5\frac{1}{4} - 2\frac{3}{5}$

2. Simplify:

a) $\frac{84}{96}$

b) $\frac{77}{42}$

c) $\frac{84}{91}$

3. Write the first quantity as a fraction of the second quantity:

a) 13 p; £1

b) 233 days; 1 leap year

4. Write either < or > between the following pairs of fractions:

a) $\frac{13}{20}$ $\frac{7}{15}$

b) $\frac{5}{9}$ $\frac{11}{18}$

c) $\frac{5}{6}$ $\frac{7}{8}$

5. Write the following fractions in ascending size order:

a) $\frac{7}{20}, \frac{3}{8}, \frac{2}{5}, \frac{3}{10}$ b) $\frac{2}{5}, \frac{7}{15}, \frac{3}{10}, \frac{1}{2}$ c) $\frac{9}{16}, \frac{3}{4}, \frac{5}{8}, \frac{17}{32}$

6. In a class of 28 children, 13 live in houses with gardens, 7 live in houses without gardens and the rest live in flats. What fraction of the children

a) do not live in houses with gardens

b) live in flats?

EXERCISE 3s **1.** Calculate:

a) $\frac{6}{7} + \frac{1}{5} - \frac{3}{4}$

b) $\frac{3}{5} + \frac{4}{9} - \frac{2}{3}$

c) $\frac{1}{2} + \frac{7}{8} - 1\frac{1}{4}$

d) $2\frac{1}{4} - 1\frac{1}{3} + 2\frac{1}{6}$

e) $4\frac{1}{5} - 5\frac{1}{2} + 1\frac{3}{10}$

f) $6\frac{1}{3} - 2\frac{4}{5} + 1\frac{7}{15}$

2. Simplify:

a) $1\frac{18}{48}$

b) $2\frac{18}{45}$

c) $\frac{10}{32}$

3. Write either > or < between the following pairs of fractions:

a) $\frac{5}{8}$ $\frac{7}{11}$

b) $\frac{4}{5}$ $\frac{5}{6}$

4. Arrange the following fractions is ascending size order:

a) $\frac{3}{4}, \frac{3}{5}, \frac{1}{2}, \frac{5}{6}$ b) $\frac{1}{2}, \frac{5}{6}, \frac{5}{9}, \frac{2}{3}$

5. Write the first quantity as a fraction of the second quantity:

a) 7 minutes; 1 hour   b) 1200 people; 3600 people

c) 76 p; £1.58

6. In a bag of potatoes there are 6 large ones, 11 small ones and 2 rotten ones. What fraction of the potatoes in the bag are

a) good ones   b) not large ones?

EXERCISE 3t   **1.** Calculate:

a) $\frac{4}{5} + \frac{2}{3} - \frac{3}{10}$ b) $\frac{7}{8} + \frac{3}{5} - \frac{17}{20}$ c) $1\frac{1}{3} + \frac{5}{6} - 2\frac{1}{12}$

d) $3\frac{1}{5} - 2\frac{1}{4} + 1\frac{1}{2}$ e) $2\frac{1}{3} - 3\frac{1}{4} + 1\frac{5}{6}$ f) $6\frac{3}{4} - 4\frac{2}{3} + 1\frac{7}{12}$

2. Simplify:

a) $4\frac{12}{32}$ b) $\frac{30}{240}$ c) $\frac{108}{42}$

3. Write the first quantity as a fraction of the second quantity:

a) 5 hours; 24 hours   b) 6 seconds; 1 minute   c) 5 months; 1 year

4. Write either > or < between the following pairs of fractions:

a) $\frac{7}{10}$ $\frac{5}{9}$ b) $\frac{2}{3}$ $\frac{5}{7}$

5. Arrange the following fractions in ascending size order:

a) $\frac{5}{11}, \frac{1}{2}, \frac{13}{22}, \frac{23}{44}$ b) $\frac{2}{3}, \frac{5}{9}, \frac{3}{4}, \frac{7}{12}$

6. A vase of flowers contains 5 pink ones, 3 red ones and 7 white ones. What fraction of the flowers in the vase are

a) red   b) not white   c) pink?

4 FRACTIONS: MULTIPLICATION AND DIVISION

MULTIPLYING FRACTIONS

When fractions are multiplied the result is given by multiplying together the numbers in the numerator and also multiplying together the numbers in the denominator. For example

$$\frac{1}{2} \times \frac{1}{3} = \frac{1 \times 1}{2 \times 3}$$

$$= \frac{1}{6}$$

If we look at a cake diagram we can see that $\frac{1}{2}$ of $\frac{1}{3}$ of the cake is $\frac{1}{6}$ of the cake.

So $\quad \dfrac{1}{2}$ of $\dfrac{1}{3} = \dfrac{1}{6}$

and $\quad \dfrac{1}{2} \times \dfrac{1}{3} = \dfrac{1}{6}$

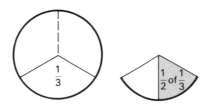

We see that "of" means "multiplied by".

EXERCISE 4a Draw cake diagrams to show that:

1. $\dfrac{1}{2} \times \dfrac{1}{4} = \dfrac{1}{8}$

2. $\dfrac{1}{3} \times \dfrac{1}{2} = \dfrac{1}{6}$

3. $\dfrac{1}{2} \times \dfrac{3}{4} = \dfrac{3}{8}$

4. $\dfrac{2}{3} \times \dfrac{1}{3} = \dfrac{2}{9}$

5. $\dfrac{1}{3} \times \dfrac{2}{5} = \dfrac{2}{15}$

6. $\dfrac{1}{4} \times \dfrac{1}{3} = \dfrac{1}{12}$

57

SIMPLIFYING

Sometimes we can simplify a product by cancelling the common factors. For example

$$\frac{2}{3} \times \frac{3}{4} = \frac{\cancel{2}^{1}}{\cancel{3}_{1}} \times \frac{\cancel{3}^{1}}{\cancel{4}_{2}} = \frac{1 \times 1}{1 \times 2}$$

$$= \frac{1}{2}$$

The diagram shows that

$$\frac{2}{3} \text{ of } \frac{3}{4} = \frac{1}{2}$$

EXERCISE 4b   Find:

$$\frac{4}{25} \times \frac{15}{16}$$

$$\frac{\cancel{4}^{1}}{\cancel{25}_{5}} \times \frac{\cancel{15}^{3}}{\cancel{16}_{4}} = \frac{1 \times 3}{5 \times 4}$$

$$= \frac{3}{20}$$

1. $\frac{3}{4} \times \frac{1}{2}$ **5.** $\frac{3}{4} \times \frac{4}{7}$ **9.** $\frac{5}{6} \times \frac{1}{4}$

2. $\frac{2}{3} \times \frac{5}{7}$ **6.** $\frac{4}{9} \times \frac{1}{7}$ **10.** $\frac{2}{3} \times \frac{7}{9}$

3. $\frac{2}{5} \times \frac{1}{3}$ **7.** $\frac{3}{7} \times \frac{2}{5}$ **11.** $\frac{3}{4} \times \frac{1}{5}$

4. $\frac{1}{2} \times \frac{7}{8}$ **8.** $\frac{2}{5} \times \frac{3}{5}$ **12.** $\frac{1}{7} \times \frac{3}{5}$

13. $\frac{7}{8} \times \frac{4}{21}$ **15.** $\frac{21}{22} \times \frac{11}{27}$ **17.** $\frac{7}{9} \times \frac{3}{21}$

14. $\frac{3}{4} \times \frac{16}{21}$ **16.** $\frac{8}{9} \times \frac{33}{44}$ **18.** $\frac{3}{4} \times \frac{5}{7}$

19. $\dfrac{4}{5} \times \dfrac{15}{16}$ **21.** $\dfrac{4}{15} \times \dfrac{25}{64}$ **23.** $\dfrac{3}{7} \times \dfrac{28}{33}$

20. $\dfrac{10}{11} \times \dfrac{33}{35}$ **22.** $\dfrac{2}{3} \times \dfrac{33}{40}$ **24.** $\dfrac{48}{55} \times \dfrac{5}{12}$

Find:

$$\dfrac{3}{5} \times \dfrac{15}{16} \times \dfrac{4}{7}$$

$$\dfrac{3}{\cancel{5}_1} \times \dfrac{\cancel{15}^3}{\cancel{16}_4} \times \dfrac{\cancel{4}^1}{7} = \dfrac{3 \times 3 \times 1}{1 \times 4 \times 7}$$

$$= \dfrac{9}{28}$$

25. $\dfrac{3}{7} \times \dfrac{5}{9} \times \dfrac{14}{15}$ **29.** $\dfrac{3}{10} \times \dfrac{5}{9} \times \dfrac{6}{7}$ **33.** $\dfrac{6}{5} \times \dfrac{4}{3} \times \dfrac{10}{4}$

26. $\dfrac{11}{21} \times \dfrac{30}{31} \times \dfrac{7}{55}$ **30.** $\dfrac{5}{7} \times \dfrac{3}{8} \times \dfrac{21}{30}$ **34.** $\dfrac{9}{8} \times \dfrac{1}{3} \times \dfrac{4}{27}$

27. $\dfrac{15}{16} \times \dfrac{8}{9} \times \dfrac{4}{5}$ **31.** $\dfrac{1}{2} \times \dfrac{7}{12} \times \dfrac{18}{35}$ **35.** $\dfrac{7}{16} \times \dfrac{9}{11} \times \dfrac{8}{21}$

28. $\dfrac{5}{6} \times \dfrac{8}{25} \times \dfrac{3}{4}$ **32.** $\dfrac{7}{11} \times \dfrac{8}{9} \times \dfrac{33}{28}$ **36.** $\dfrac{5}{14} \times \dfrac{21}{25} \times \dfrac{5}{9}$

MULTIPLYING MIXED NUMBERS

Suppose that we want to find the value of $2\frac{1}{3} \times \frac{5}{21} \times 1\frac{1}{5}$.

We cannot multiply mixed numbers together unless we change them into improper fractions first. So we change $2\frac{1}{3}$ into $\frac{7}{3}$ and we change $1\frac{1}{5}$ into $\frac{6}{5}$. Then we can use the same method as before.

EXERCISE 4c

Find $2\frac{1}{3} \times \frac{5}{21} \times 1\frac{1}{5}$

$$2\frac{1}{3} \times \frac{5}{21} \times 1\frac{1}{5} = \dfrac{\cancel{7}^1}{\cancel{3}_1} \times \dfrac{\cancel{5}^1}{\cancel{21}_3} \times \dfrac{\cancel{6}^2}{\cancel{5}_1}$$

$$= \dfrac{2}{3}$$

Find:

1. $1\frac{1}{2} \times \frac{2}{5}$

2. $2\frac{1}{2} \times \frac{4}{5}$

3. $3\frac{1}{4} \times \frac{3}{13}$

4. $4\frac{2}{3} \times 2\frac{2}{5}$

5. $2\frac{1}{5} \times \frac{5}{22}$

6. $1\frac{1}{4} \times \frac{2}{5}$

7. $2\frac{1}{3} \times \frac{3}{8}$

8. $\frac{10}{11} \times 2\frac{1}{5}$

9. $3\frac{1}{2} \times 4\frac{2}{3}$

10. $4\frac{1}{4} \times \frac{4}{21}$

11. $5\frac{1}{4} \times 2\frac{2}{3}$

12. $3\frac{5}{7} \times 1\frac{1}{13}$

13. $8\frac{1}{3} \times 3\frac{3}{5}$

14. $2\frac{1}{10} \times 7\frac{6}{7}$

15. $6\frac{3}{10} \times 1\frac{4}{21}$

16. $4\frac{2}{7} \times 2\frac{1}{10}$

17. $6\frac{1}{4} \times 1\frac{3}{5}$

18. $5\frac{1}{2} \times 1\frac{9}{11}$

19. $8\frac{3}{4} \times 2\frac{2}{7}$

20. $16\frac{1}{2} \times 3\frac{7}{11}$

21. $6\frac{2}{5} \times 1\frac{7}{8} \times \frac{7}{12}$

22. $2\frac{4}{7} \times 4\frac{2}{3} \times 1\frac{1}{4}$

23. $3\frac{2}{3} \times 1\frac{1}{5} \times 1\frac{3}{22}$

24. $1\frac{1}{18} \times 1\frac{4}{5} \times 3\frac{1}{3}$

25. $4\frac{4}{5} \times 1\frac{5}{18} \times 3\frac{3}{4}$

26. $7\frac{1}{2} \times 1\frac{1}{3} \times \frac{9}{10}$

27. $3\frac{1}{5} \times 2\frac{1}{2} \times 1\frac{3}{4}$

28. $4\frac{1}{2} \times 1\frac{1}{7} \times 2\frac{1}{3}$

29. $2\frac{1}{3} \times \frac{6}{11} \times 2\frac{5}{14}$

30. $3\frac{9}{10} \times 1\frac{2}{3} \times 1\frac{3}{13}$

WHOLE NUMBERS AS FRACTIONS

A whole number can be written as a fraction with a denominator of 1. For instance $6 = \frac{6}{1}$.

Doing this makes it easier to multiply a whole number by a fraction or a mixed number.

EXERCISE 4d Find:

$$6 \times 7\tfrac{1}{3}$$

$$6 \times 7\tfrac{1}{3} = \tfrac{\overset{2}{6}}{1} \times \tfrac{22}{\underset{1}{3}}$$
$$= 44$$

1. $5 \times 4\tfrac{3}{5}$ **5.** $18 \times 6\tfrac{1}{9}$ **9.** $5\tfrac{5}{7} \times 21$

2. $2\tfrac{1}{7} \times 14$ **6.** $4 \times 3\tfrac{3}{8}$ **10.** $3 \times 6\tfrac{1}{9}$

3. $3\tfrac{1}{8} \times 4$ **7.** $3\tfrac{3}{5} \times 10$ **11.** $1\tfrac{3}{4} \times 8$

4. $4\tfrac{1}{6} \times 9$ **8.** $2\tfrac{5}{6} \times 3$ **12.** $28 \times 1\tfrac{4}{7}$

FRACTIONS OF QUANTITIES

EXERCISE 4e

Find three fifths of 95 metres.

$$\tfrac{3}{\underset{1}{5}} \times \tfrac{\overset{19}{95}}{1} = 57$$

$\tfrac{3}{5}$ of 95 metres is 57 metres.

Find three quarters of £1.

$$£1 = 100 \text{ pence}$$
$$\tfrac{3}{\underset{1}{4}} \times \tfrac{\overset{25}{100}}{1} = 75$$

$\tfrac{3}{4}$ of £1 is 75 pence.

Find:

1. $\tfrac{1}{3}$ of 18   **6.** $\tfrac{1}{4}$ of 24

2. $\tfrac{1}{5}$ of 30   **7.** $\tfrac{1}{6}$ of 30

3. $\tfrac{1}{7}$ of 21   **8.** $\tfrac{1}{8}$ of 64

4. $\tfrac{2}{3}$ of 24   **9.** $\tfrac{5}{6}$ of 36

5. $\tfrac{5}{7}$ of 14   **10.** $\tfrac{3}{8}$ of 40

11. $\frac{3}{5}$ of 20 metres

12. $\frac{5}{9}$ of 45 dollars

13. $\frac{9}{10}$ of 50 litres

14. $\frac{3}{8}$ of 88 miles

15. $\frac{7}{16}$ of 48 gallons

16. $\frac{4}{9}$ of 18 metres

17. $\frac{5}{8}$ of 16 dollars

18. $\frac{4}{9}$ of 63 litres

19. $\frac{3}{7}$ of 35 miles

20. $\frac{8}{11}$ of 121 gallons

21. $\frac{1}{4}$ of £2

22. $\frac{2}{9}$ of 36 pence

23. $\frac{3}{10}$ of £1

24. $\frac{2}{7}$ of 42 pence

25. $\frac{4}{5}$ of 1 year (365 days)

26. $\frac{3}{8}$ of 1 day (24 hours)

27. $\frac{1}{7}$ of 1 week

28. $\frac{1}{3}$ of £9

29. $\frac{3}{5}$ of £1

30. $\frac{7}{8}$ of 1 day (24 hours)

DIVIDING BY FRACTIONS

When we divide 6 by 3 we are finding how many threes there are in 6 and we say $6 \div 3 = 2$.

In the same way, when we divide 10 by $\frac{1}{2}$ we are finding how many halves there are in 10; we know that there are 20, so we say $10 \div \frac{1}{2} = 20$.

But we also know that $10 \times 2 = 20$ so

$$\frac{10}{1} \div \frac{1}{2} = 20 \qquad \text{and} \qquad \frac{10}{1} \times \frac{2}{1} = 20$$

This example shows that $\frac{10}{1} \div \frac{1}{2} = \frac{10}{1} \times \frac{2}{1}$.

To divide by a fraction we turn that fraction upside down and multiply.

EXERCISE 4f

How many thirds are there in 5?

$$5 \div \frac{1}{3} = 5 \times \frac{3}{1}$$

$$= \frac{5}{1} \times \frac{3}{1}$$

$$= 15$$

There are 15 thirds in 5.

Find the value of $36 \div \frac{3}{4}$.

$$\frac{36}{1} \div \frac{3}{4} = \frac{\overset{12}{\cancel{36}}}{1} \times \frac{4}{\underset{1}{\cancel{3}}}$$

$$= \frac{48}{1}$$

$$= 48$$

Divide $\frac{7}{16}$ by $\frac{5}{8}$.

$$\frac{7}{16} \div \frac{5}{8} = \frac{7}{\underset{2}{\cancel{16}}} \times \frac{\overset{1}{\cancel{8}}}{5}$$

$$= \frac{7}{10}$$

1. How many $\frac{1}{2}$s are there in 7?

2. How many $\frac{1}{4}$s are there in 5?

3. How many times does $\frac{1}{7}$ go into 3?

4. How many $\frac{3}{5}$s are there in 9?

5. How many times does $\frac{2}{3}$ go into 8?

Find:

6. $8 \div \frac{4}{5}$

7. $18 \div \frac{6}{7}$

8. $40 \div \frac{8}{9}$

9. $72 \div \frac{8}{11}$

10. $28 \div \frac{14}{15}$

11. $15 \div \frac{5}{6}$

12. $14 \div \frac{7}{8}$

13. $35 \div \frac{5}{7}$

14. $44 \div \frac{4}{9}$

15. $27 \div \frac{9}{13}$

16. $36 \div \frac{4}{7}$

17. $34 \div \frac{17}{19}$

18. $\dfrac{21}{32} \div \dfrac{7}{8}$ **20.** $\dfrac{3}{56} \div \dfrac{9}{14}$ **22.** $\dfrac{8}{75} \div \dfrac{4}{15}$

19. $\dfrac{9}{25} \div \dfrac{3}{10}$ **21.** $\dfrac{21}{22} \div \dfrac{7}{11}$ **23.** $\dfrac{35}{42} \div \dfrac{5}{6}$

24. $\dfrac{28}{27} \div \dfrac{4}{9}$ **26.** $\dfrac{15}{26} \div \dfrac{5}{13}$ **28.** $\dfrac{8}{21} \div \dfrac{4}{7}$

25. $\dfrac{22}{45} \div \dfrac{11}{15}$ **27.** $\dfrac{49}{50} \div \dfrac{7}{10}$ **29.** $\dfrac{9}{26} \div \dfrac{12}{13}$

DIVIDING BY WHOLE NUMBERS AND MIXED NUMBERS

If we want to divide 3 by 5 we can say

$$3 \div 5 = \frac{3}{1} \div \frac{5}{1}$$

$$= \frac{3}{1} \times \frac{1}{5}$$

$$= \frac{3}{5}$$

So $3 \div 5$ is the same as $\frac{3}{5}$.

Division with mixed numbers can be done as long as all the mixed numbers are first changed into improper fractions. For example if we want to divide $3\frac{1}{8}$ by $8\frac{3}{4}$ we first change $3\frac{1}{8}$ into $\frac{25}{8}$ and $8\frac{3}{4}$ into $\frac{35}{4}$. Then we can use the same method as before.

EXERCISE 4g

Write 7 divided by 11 as a fraction.

$$7 \div 11 = \tfrac{7}{11}$$

Find the value of $3\frac{1}{8} \div 8\frac{3}{4}$.

$$3\frac{1}{8} \div 8\frac{3}{4} = \frac{25}{8} \div \frac{35}{4}$$

$$= \frac{\overset{5}{\cancel{25}}}{\underset{2}{\cancel{8}}} \times \frac{\overset{1}{\cancel{4}}}{\underset{7}{\cancel{35}}}$$

$$= \frac{5}{14}$$

Find:

1. $5\frac{4}{9} \div \frac{14}{27}$

2. $3\frac{1}{8} \div 3\frac{3}{4}$

3. $7\frac{1}{5} \div 1\frac{7}{20}$

4. Divide $8\frac{1}{4}$ by $1\frac{3}{8}$

5. Divide $6\frac{2}{3}$ by $2\frac{4}{9}$

6. $4\frac{2}{7} \div \frac{9}{14}$

7. $5\frac{5}{8} \div 6\frac{1}{4}$

8. $6\frac{4}{9} \div 1\frac{1}{3}$

9. Divide $5\frac{1}{4}$ by $2\frac{11}{12}$

10. Divide $7\frac{1}{7}$ by $1\frac{11}{14}$

11. $10\frac{2}{3} \div 1\frac{7}{9}$

12. $8\frac{4}{5} \div 3\frac{3}{10}$

13. Divide $11\frac{1}{4}$ by $\frac{15}{16}$

14. Divide $9\frac{1}{7}$ by $1\frac{11}{21}$

15. $31\frac{1}{2} \div 5\frac{5}{8}$

16. $9\frac{3}{4} \div 1\frac{5}{8}$

17. $12\frac{1}{2} \div 8\frac{3}{4}$

18. Divide $10\frac{5}{6}$ by $3\frac{1}{4}$

19. Divide $8\frac{2}{3}$ by $5\frac{7}{9}$

20. $22\frac{2}{3} \div 1\frac{8}{9}$

MIXED MULTIPLICATION AND DIVISION

Suppose we want to find the value of an expression like $2\frac{1}{4} \times \frac{3}{14} \div 1\frac{2}{7}$. Two things need to be done:

Step one—If there are any mixed numbers, change them into improper fractions.

Step two—Turn the fraction *after* the $\div$ sign upside down and change $\div$ into $\times$.

Then

$$2\frac{1}{4} \times \frac{3}{14} \div 1\frac{2}{7} \quad = \quad \frac{9}{4} \times \frac{3}{14} \div \frac{9}{7} \quad \text{(Step one)}$$

$$= \quad \frac{\cancel{9}^1}{4} \times \frac{3}{\cancel{14}_2} \times \frac{\cancel{7}^1}{\cancel{9}_1} \quad \text{(Step two)}$$

$$= \quad \frac{3}{8}$$

EXERCISE 4h   Find:

$$4\frac{1}{3} \times 1\frac{1}{8} \div 2\frac{1}{4}$$

$$4\frac{1}{3} \times 1\frac{1}{8} \div 2\frac{1}{4} = \frac{13}{3} \times \frac{9}{8} \div \frac{9}{4}$$

$$= \frac{13}{3} \times \frac{9^1}{8_2} \times \frac{4^1}{9_1}$$

$$= \frac{13}{6}$$

$$= 2\frac{1}{6}$$

1. $\frac{5}{8} \times 1\frac{1}{2} \div \frac{15}{16}$

5. $\frac{2}{5} \times \frac{9}{10} \div \frac{27}{40}$

9. $\frac{3}{5} \times \frac{9}{11} \div \frac{18}{55}$

2. $2\frac{3}{4} \times \frac{5}{6} \div \frac{11}{12}$

6. $\frac{3}{4} \times 2\frac{1}{3} \div \frac{21}{32}$

10. $\frac{1}{4} \times \frac{11}{12} \div \frac{22}{27}$

3. $\frac{2}{3} \times 1\frac{1}{5} \div \frac{12}{25}$

7. $3\frac{2}{5} \times \frac{4}{5} \div \frac{8}{15}$

11. $\frac{3}{7} \times \frac{2}{5} \div \frac{8}{21}$

4. $\frac{4}{7} \times \frac{8}{9} \div \frac{16}{21}$

8. $\frac{3}{7} \times 2\frac{1}{2} \div \frac{10}{21}$

12. $\frac{14}{25} \times \frac{5}{9} \div \frac{7}{18}$

MIXED OPERATIONS

When brackets are placed round a pair of fractions it means that we have to work out what is *inside* the brackets before doing anything else. For example

$$\left(\frac{1}{2}+\frac{1}{4}\right) \times \frac{5}{7} = \left(\frac{2+1}{4}\right) \times \frac{5}{7}$$

$$= \frac{3}{4} \times \frac{5}{7}$$

$$= \frac{15}{28}$$

If we meet an expression in which $+$, $-$, $\times$ and $\div$ occur, we need to know the order in which to do the calculations. We use the same rule for fractions as we used for whole numbers, that is

Brackets first, then Multiply and Divide, then Add and Subtract.

You may remember this order from the phrase
"Bless My Dear Aunt Sally".

EXERCISE 4i   Calculate:

$$\frac{7}{8} \div \left(1\frac{1}{2} \times 1\frac{5}{9}\right)$$

$$\frac{7}{8} \div \left(1\frac{1}{2} \times 1\frac{5}{9}\right) = \frac{7}{8} \div \left(\frac{3}{2} \times \frac{14}{9}\right)$$

$$= \frac{7}{8} \div \frac{7}{3} \qquad \text{(brackets)}$$

$$= \frac{7}{8} \times \frac{3}{7} \qquad \text{(divide)}$$

$$= \frac{3}{8}$$

$$\frac{2}{3} \times \left(\frac{1}{4} - \frac{1}{12}\right) \div \frac{1}{2}$$

$$\frac{2}{3} \times \left(\frac{1}{4} - \frac{1}{12}\right) \div \frac{1}{2} = \frac{2}{3} \times \left(\frac{3-1}{12}\right) \div \frac{1}{2}$$

$$= \frac{2}{3} \times \frac{2}{12} \div \frac{1}{2} \qquad \text{(brackets)}$$

$$= \frac{2}{3} \times \frac{2}{12} \times \frac{2}{1} \qquad \text{(multiply and divide)}$$

$$= \frac{2}{9}$$

$$\frac{2}{5} - \frac{1}{2} \times \frac{3}{5} + \frac{1}{10}$$

$$\frac{2}{5} - \frac{1}{2} \times \frac{3}{5} + \frac{1}{10} = \frac{2}{5} - \frac{3}{10} + \frac{1}{10} \qquad \text{(multiply)}$$

$$= \frac{4-3+1}{10} \qquad \text{(add and subtract)}$$

$$= \frac{2}{10}$$

$$= \frac{1}{5}$$

1. $\frac{1}{2} + \frac{1}{4} \times \frac{2}{5}$

2. $\frac{2}{3} \times \frac{1}{2} + \frac{1}{4}$

3. $\frac{4}{5} - \frac{3}{10} \div \frac{1}{2}$

4. $\frac{2}{7} \div \frac{2}{3} - \frac{3}{14}$

5. $\frac{4}{5} + \frac{3}{10} \times \frac{2}{9}$

6. $\frac{1}{3} - \frac{1}{2} \times \frac{1}{4}$

7. $\frac{3}{4} \div \frac{1}{2} + \frac{1}{8}$

8. $\frac{1}{7} + \frac{5}{8} \div \frac{3}{4}$

9. $\frac{5}{6} \times \frac{3}{10} - \frac{3}{16}$

10. $\frac{3}{7} - \frac{1}{4} \times \frac{8}{21}$

11. $\left(\frac{4}{9} - \frac{1}{3}\right) \times \frac{6}{7}$

16. $\frac{3}{8} \div \left(\frac{2}{3} + \frac{1}{4}\right)$

12. $\frac{3}{5} \times \left(\frac{2}{3} + \frac{1}{2}\right)$

17. $\left(\frac{4}{7} + \frac{1}{3}\right) \div 3\frac{4}{5}$

13. $\frac{7}{8} \div \left(\frac{3}{4} + \frac{2}{3}\right)$

18. $\frac{5}{9} \times \left(\frac{2}{3} - \frac{1}{6}\right)$

14. $\left(\frac{3}{10} + \frac{2}{5}\right) \div \frac{7}{15}$

19. $\left(\frac{6}{11} - \frac{1}{2}\right) \div \frac{3}{4}$

15. $\left(\frac{5}{11} - \frac{1}{3}\right) \times \frac{3}{8}$

20. $\frac{9}{10} \div \left(\frac{1}{6} + \frac{2}{3}\right)$

21. $\frac{1}{6} \times \left(\frac{2}{3} - \frac{1}{2}\right) \div \frac{7}{12}$

26. $\frac{2}{9} + \left(\frac{6}{7} \div \frac{3}{4}\right) \times 3\frac{1}{2}$

22. $\frac{7}{10} \div \left(\frac{2}{5} + \frac{4}{15} \times \frac{3}{5}\right)$

27. $1\frac{1}{10} \times \frac{23}{24} \div \left(\frac{3}{5} + \frac{1}{6}\right)$

23. $\left(2\frac{1}{4} + \frac{3}{8}\right) \times \frac{2}{3} - 1\frac{1}{2}$

28. $2\frac{2}{5} - \frac{7}{10} \times \left(\frac{4}{7} - \frac{1}{3}\right)$

24. $1\frac{3}{11} - \frac{6}{7} \times 1\frac{5}{9} + \frac{13}{33}$

29. $\frac{5}{9} \div \left(1\frac{1}{3} + \frac{4}{9}\right) + \frac{3}{8}$

25. $\frac{5}{8} \times \left(\frac{4}{9} - \frac{1}{6}\right) \div 1\frac{9}{16}$

30. $1\frac{2}{9} + \left(\frac{5}{6} - \frac{3}{4} \div 4\frac{1}{2}\right)$

State whether each of the following statements is true or false:

31. $\frac{1}{2} \times \frac{2}{3} + \frac{1}{3} = \frac{1}{3} + \frac{1}{3}$

36. $\frac{3}{4} - \frac{1}{2} \times \frac{2}{3} = \frac{1}{4} \times \frac{2}{3}$

32. $\frac{1}{3} \times \frac{3}{4} + \frac{1}{4} = \frac{1}{3} \times 1$

37. $\frac{2}{3} - \frac{1}{4} + \frac{1}{2} = \frac{2}{3} + \frac{1}{2} - \frac{1}{4}$

33. $\frac{1}{4} \div \frac{3}{4} + \frac{1}{2} = \frac{1}{3} + \frac{1}{2}$

38. $\frac{3}{5} \times \frac{2}{3} + \frac{1}{2} = \frac{3}{5} + \frac{1}{2} \times \frac{2}{3}$

34. $\frac{1}{3} + \frac{2}{3} \times \frac{1}{4} = \frac{1}{3} + \frac{1}{6}$

39. $\frac{4}{7} - \frac{1}{4} \div \frac{1}{3} = \left(\frac{4}{7} - \frac{1}{4}\right) \div \frac{1}{3}$

35. $\frac{1}{2} + \frac{1}{4} \div \frac{1}{2} = \frac{3}{4} \times \frac{2}{1}$

40. $\frac{3}{8} \div \frac{1}{4} - \frac{1}{4} = \frac{3}{8} \times \frac{4}{1} - \frac{1}{4}$

EXERCISE 4j   In this exercise you will find $+$, $-$, $\times$ and $\div$. Read the question carefully and then decide which method to use. Find:

1. $1\frac{1}{2} + 3\frac{1}{4}$

6. $4\frac{1}{4} \times \frac{2}{9}$

2. $2\frac{3}{8} - 1\frac{1}{4}$

7. $3\frac{2}{3} \div \frac{1}{6}$

3. $1\frac{1}{5} \times \frac{5}{8}$

8. $2\frac{1}{5} - 1\frac{1}{3}$

4. $3\frac{1}{2} \div \frac{7}{8}$

9. $5\frac{1}{2} \times \frac{6}{11}$

5. $\frac{4}{7} + 1\frac{1}{2}$

10. $1\frac{3}{8} + 2\frac{1}{2}$

11. $5\frac{1}{2} + \frac{3}{4}$

12. $4\frac{1}{3} \times \frac{6}{13}$

13. $3\frac{4}{5} - 2\frac{1}{10}$

14. $3\frac{1}{7} \div 1\frac{3}{8}$

15. $4\frac{1}{5} \times \frac{4}{7}$

16. $2\frac{5}{6} \div 3\frac{1}{3}$

17. $1\frac{4}{7} + 2\frac{1}{2}$

18. $2\frac{3}{4} - 1\frac{7}{8}$

19. $2\frac{3}{8} + 1\frac{7}{16}$

20. $5\frac{1}{4} \div 1\frac{1}{6}$

21. $1\frac{1}{4} + \frac{2}{3} - \frac{5}{6}$

22. $2\frac{1}{2} - \frac{2}{3} - 1\frac{1}{4}$

23. $3\frac{1}{2} + 1\frac{1}{4} + \frac{5}{8}$

24. $2\frac{1}{3} + 1\frac{1}{2} - \frac{3}{4}$

25. $4\frac{1}{8} - 5\frac{3}{4} + 2\frac{1}{2}$

26. $4\frac{1}{2} - 5\frac{1}{4} + 2\frac{1}{8}$

27. $3\frac{4}{5} + \frac{3}{10} - 1\frac{1}{20}$

28. $5\frac{1}{2} - 1\frac{3}{4} - 2\frac{1}{4}$

29. $3\frac{1}{7} + 2\frac{1}{2} - \frac{3}{14}$

30. $5\frac{1}{2} - \frac{3}{4} - 4\frac{1}{4}$

31. $1\frac{1}{2} + 2\frac{2}{3} \times \frac{3}{4}$

32. $2\frac{1}{3} \times 1\frac{1}{2} - 2\frac{1}{2}$

33. $4\frac{1}{3} \div 2\frac{1}{6} + \frac{1}{4}$

34. $2\frac{2}{5} - \frac{6}{7} \div \frac{5}{14}$

35. $1\frac{1}{8} \times \frac{4}{9} \div 2\frac{1}{2}$

36. $2\frac{3}{8} - 1\frac{1}{5} \times 1\frac{2}{3}$

37. $1\frac{3}{4} \div 4\frac{2}{3} - \frac{5}{16}$

38. $2\frac{1}{7} \times 3\frac{1}{4} \div 1\frac{5}{8}$

39. $1\frac{1}{2} + 2\frac{5}{7} - 1\frac{5}{14}$

40. $\frac{3}{5} \times 1\frac{1}{4} \div \frac{3}{8}$

PROBLEMS

EXERCISE 4k

If Jane can iron a shirt in $4\frac{3}{4}$ minutes, how long will it take her to iron 10 shirts?

Time to iron 1 shirt $= 4\frac{3}{4}$ minutes

Time to iron 10 shirts $= 4\frac{3}{4} \times 10$ minutes

$= \frac{19}{4} \times \frac{10}{1}$ minutes

$= \frac{95}{2}$ minutes

$= 47\frac{1}{2}$ minutes

A piece of string of length $22\frac{1}{2}$ cm is to be cut into small pieces each $\frac{3}{4}$ cm long. How many pieces can be obtained?

Number of small pieces
= length of string ÷ length of one short piece

$$= 22\frac{1}{2} \div \frac{3}{4}$$

$$= \frac{\overset{15}{\cancel{45}}}{\underset{1}{\cancel{2}}} \times \frac{\overset{2}{\cancel{4}}}{\underset{1}{\cancel{3}}}$$

$$= 30$$

Thus 30 pieces can be obtained.

1. A bag of flour weighs $1\frac{1}{2}$ kilograms. What is the weight of 20 bags?

2. A cook adds $3\frac{1}{2}$ cups of water to a stew. If the cup holds $\frac{1}{10}$ of a litre how many litres of water were added?

3. My journey to school starts with a walk of $\frac{1}{2}$ km to the bus stop, then a bus ride of $2\frac{1}{5}$ km followed by a walk of $\frac{3}{10}$ km. How long is my journey to school?

4. It takes $3\frac{1}{4}$ minutes for a cub scout to clean a pair of shoes. If he cleans 18 pairs of shoes to raise money for a charity, how long does he spend on the job?

5. A burger bar chef cooks some beefburgers and piles them one on top of the other. If each burger is $9\frac{1}{2}$ mm thick and the pile is 209 mm high, how many did he cook?

6. If you read 30 pages of a book in $\frac{3}{4}$ of an hour, how many minutes does it take to read each page?

MIXED EXERCISES

EXERCISE 4I   **1.** Calculate

a) $\frac{3}{4} + \frac{11}{12}$ b) $3\frac{1}{8} - 2\frac{1}{4} + 1\frac{1}{2}$

2. Find how many times $2\frac{1}{4}$ goes into $13\frac{1}{2}$.

3. What is $\frac{7}{9}$ of $1\frac{1}{14}$?

4. Find $\frac{3}{5} + 1\frac{1}{2} \times \frac{7}{10}$.

5. Arrange the following fractions in ascending order of size: $\frac{7}{10}, \frac{3}{5}, \frac{2}{3}$.

6. Find:

    a) $4\frac{1}{7} \times 4\frac{2}{3}$ b) $3\frac{3}{8} \div 2\frac{1}{4}$.

7. What is $\frac{3}{4}$ of $\frac{8}{9}$ added to $1\frac{1}{2}$?

8. Find $\left(1\frac{7}{8} + 2\frac{1}{4}\right) \times 1\frac{5}{11}$.

9. What is $\frac{2}{7}$ of 1 hour 3 minutes (in minutes)?

10. Find $7\frac{1}{5} - 4\frac{1}{8} \div 1\frac{1}{4}$.

11. Fill in the missing numbers:

    a) $\frac{7}{9} = \frac{21}{\underline{}}$ b) $\frac{10}{11} = \frac{\underline{}}{44}$

12. Express as mixed numbers:

    a) $\frac{13}{5}$ b) $\frac{31}{8}$ c) $\frac{27}{5}$

13. State whether the following statements are true or false:

    a) $\frac{4}{11} > \frac{3}{10}$ b) $\frac{3}{7}$ of $5 = \frac{3}{7} \times \frac{5}{1}$ c) $2\frac{1}{7} = \frac{7}{15}$

14. A handyman takes $1\frac{1}{8}$ minutes to lay one brick. How long will it take him to lay 56 bricks?

15. A pharmacist counts 48 tablets and puts them in a bottle. Each tablet weighs $\frac{1}{4}$ of a gram and the weight of the empty bottle is $112\frac{1}{2}$ grams. What is the total weight?

EXERCISE 4m   **1.** Find:

    a) $4\frac{1}{2} \times 3\frac{1}{3}$ b) $3\frac{2}{5} \div \frac{3}{10}$

2. Find:

    a) $\frac{8}{9} + \frac{21}{27}$ b) $2\frac{1}{3} + \frac{4}{9} + 1\frac{5}{6}$

3. Put $>$ or $<$ between the following pairs of numbers:

    a) $\frac{4}{7}$ $\frac{5}{8}$ b) $\frac{11}{9}$ $1\frac{3}{10}$

4. Calculate:

    a) $5\frac{1}{4} - 1\frac{2}{3} \div \frac{2}{5}$ b) $3\frac{3}{8} \times \left(8\frac{1}{2} - 5\frac{5}{6}\right)$

5. Arrange in ascending order: $\frac{7}{15}$, $\frac{1}{3}$, $\frac{2}{5}$.

6. What is $1\frac{1}{2}$ subtracted from $\frac{2}{3}$ of $5\frac{1}{4}$?

7. Find:

a) $4\frac{1}{2} \times 3\frac{2}{3} - 10\frac{1}{4}$ b) $3\frac{1}{2} \div \left(2\frac{1}{8} - \frac{3}{4}\right)$

8. What is $1\frac{2}{3}$ of 1 minute 15 seconds (in seconds)?

9. Fill in the missing numbers:

a) $\frac{4}{5} = \frac{}{30}$ b) $\frac{2}{7} = \frac{6}{}$

10. Express as mixed numbers:

a) $\frac{25}{8}$ b) $\frac{49}{9}$ c) $\frac{37}{6}$

11. A car travels $5\frac{1}{4}$ km north, then $2\frac{1}{2}$ km west and finally $4\frac{3}{8}$ km north. What is the total distance travelled (in kilometres)? What fraction of the journey was travelled in a northerly direction?

12. A man can paint a door in 1 hour 15 minutes. How many similar doors can he paint in $7\frac{1}{2}$ hours?

EXERCISE 4n

1. Find:

a) $1\frac{5}{6} + \frac{5}{18} + \frac{7}{12}$ b) $1\frac{2}{3} - 2\frac{1}{5} + \frac{8}{15}$

2. Find:

a) $1\frac{5}{6} \div 7\frac{1}{3}$ b) $2\frac{1}{4} \times \frac{16}{45}$

3. What is $\frac{5}{6}$ of the number of days in June?

4. Arrange in descending order: $\frac{17}{20}$, $\frac{3}{4}$, $\frac{7}{10}$

5. Calculate:

a) $4\frac{1}{2} \times 3\frac{2}{3} - 10\frac{1}{4}$ b) $3\frac{1}{4} + 5\frac{1}{2} \div \frac{3}{8}$

6. Which is smaller $\frac{8}{11}$ or $\frac{7}{9}$?

7. Find $3\frac{9}{10} \div \left(3\frac{3}{5} - 1\frac{1}{2}\right)$.

8. What is $\frac{4}{7}$ of $4\frac{2}{3}$ divided by $1\frac{1}{9}$?

9. Express as mixed numbers:

a) $\frac{22}{3}$ b) $\frac{46}{5}$ c) $\frac{106}{10}$

10. Which of the following statements are true?

a) $3\frac{1}{2} \div 1 = 3\frac{1}{2} \times 1$ b) $\frac{1}{2} \times \left(\frac{1}{4} + \frac{1}{8}\right) =$ half of $\frac{3}{8}$

c) $\frac{1}{2} + \frac{1}{2} \div 2 = \frac{3}{4}$

11. It takes $1\frac{3}{4}$ minutes to wrap a parcel and a half a minute to address it. How long does it take to wrap and address 8 similar parcels?

12. My bag contains 2 books each of weight $\frac{3}{7}$ kg and 3 folders each of weight $\frac{5}{21}$ kg. What is the total weight in my bag? What fraction of the total weight is books?

5 INTRODUCTION TO DECIMALS

THE MEANING OF DECIMALS

Consider the number 426. The position of the figures indicates what each figure represents. We can write:

hundreds   tens   units

4   2   6

Each quantity in the heading is $\frac{1}{10}$ of the quantity to its left: ten is $\frac{1}{10}$ of a hundred, a unit is $\frac{1}{10}$ of ten. Moving further to the right we can have further headings: tenths of a unit, hundredths of a unit and so on (a hundredth of a unit is $\frac{1}{10}$ of a tenth of a unit). For example:

tens   units   tenths   hundredths

1   6   .   0   2

To mark where the units come we put a point after the units position. 16.02 is 1 ten, 6 units and 2 hundredths or $16\frac{2}{100}$.

units   tenths   hundredths   thousandths

0   .   0   0   4

0.004 is 4 thousandths or $\frac{4}{1000}$. In this case, 0 is written before the point to help make it clear where the point comes.

EXERCISE 5a   Write the following numbers in headed columns:

		tens	units	tenths	hundredths
34.62	=	3	4 .	6	2

		units	tenths	hundredths	thousandths	ten-thousandths
0.0207	=	0 .	0	2	0	7

1.	2.6	**5.**	101.3	**9.**	6.34
2.	32.1	**6.**	0.00007	**10.**	0.604
3.	6.03	**7.**	1.046	**11.**	15.045
4.	0.09	**8.**	12.001	**12.**	0.0092

CHANGING DECIMALS TO FRACTIONS

EXERCISE 5b   Write the following decimals as fractions in their lowest terms (using mixed numbers where necessary):

0.6

$$\text{units} \quad \text{tenths}$$

$$0.6 \;=\; 0 \;.\; 6 \;=\; \frac{6}{10}$$

$$=\; \frac{3}{5}$$

12.04

$$\text{tens} \quad \text{units} \quad \text{tenths} \quad \text{hundredths}$$

$$12.04 \;=\; 1 \quad 2 \;.\; 0 \quad\quad 4 \;=\; 12\frac{4}{100}$$

$$=\; 12\frac{1}{25}$$

---

1. 0.2

2. 0.06

3. 1.3

4. 0.0007

5. 0.001

6. 6.4

7. 0.7

8. 2.01

9. 1.8

10. 1.7

11. 15.5

12. 8.06

0.21

$$\text{units} \quad \text{tenths} \quad \text{hundredths}$$

$$0.21 \;=\; 0 \;.\; 2 \quad\quad 1 \quad = \frac{2}{10} + \frac{1}{100}$$

$$=\; \frac{20+1}{100}$$

$$=\; \frac{21}{100}$$

0.403

	units	tenths	hundredths	thousandths
0.403 =	0 .	4	0	3

$$= \frac{4}{10} + \frac{3}{1000}$$

$$= \frac{400 + 3}{1000}$$

$$= \frac{403}{1000}$$

You can go straight from the decimal to one fraction.

0.302

	units	tenths	hundredths	thousandths	
0.302 =	0 .	3	0	2	$= \frac{302}{1000}$
					$= \frac{151}{500}$

Write as fractions:

13.	0.73	**17.**	0.000 67	**21.**	0.0207
14.	0.081	**18.**	0.17	**22.**	0.63
15.	0.207	**19.**	0.071	**23.**	0.031
16.	0.0029	**20.**	0.3001	**24.**	0.47

Write as fractions in their lowest terms:

25.	0.25	**29.**	0.15	**33.**	0.044
26.	0.072	**30.**	0.025	**34.**	0.125
27.	0.38	**31.**	0.35	**35.**	0.48
28.	0.0305	**32.**	0.0016	**36.**	0.625

CHANGING FRACTIONS TO DECIMALS

EXERCISE 5c Write the following numbers as decimals:

$\frac{7}{10}$

units tenths

$\frac{7}{10}$ = 0 . 7

$3\frac{3}{100}$

units tenths hundredths

$3\frac{3}{100}$ = 3 . 0   3

1. $\frac{3}{100}$ **5.** $\frac{4}{10}$ **9.** $7\frac{8}{100}$

2. $\frac{9}{10}$ **6.** $2\frac{6}{100}$ **10.** $\frac{6}{10\,000}$

3. $1\frac{1}{10}$ **7.** $\frac{4}{100}$ **11.** $4\frac{5}{1000}$

4. $\frac{2}{1000}$ **8.** $7\frac{8}{10}$ **12.** $\frac{29}{10\,000}$

ADDITION OF DECIMALS

To add decimals we add in the usual way.

	tens	units	tenths	
$4.2 + 13.1 = 17.3$		4 .	2	2 tenths + 1 tenth
+	1	3 .	1	= 3 tenths
	1	7 .	3	

	units	tenths	
$5.3 + 6.8 = 12.1$	5 .	3	3 tenths + 8 tenths
+	6 .	8	= 11 tenths
	1 2 .	1	= 1 unit and 1 tenth

The headings above the figures need not be written as long as we know what they are and the decimal points are in line (including the invisible point after a whole number, e.g. $4 = 4.0$).

EXERCISE 5d   Find:

32.6 + 1.7	32.6
	+ 1.7
	34.3
32.6 + 1.7 = 34.3	

3 + 1.6 + 0.032 + 2.0066	3
	1.6
	0.032
3 + 1.6 + 0.032 + 2.0066 = 6.6386	+ 2.0066
	6.6386

1. 7.2 + 3.6

2. 6.21 + 1.34

3. 0.013 + 0.026

4. 3.87 + 0.11

5. 4.6 + 1.23

6. 13.14 + 0.9

7. 4 + 3.6

8. 9.24 + 3

9. 3.6 + 0.08

10. 7.2 + 0.32 + 1.6

11. 0.0043 + 0.263

12. 0.002 + 2.1

13. 0.00052 + 0.00124

14. 0.068 + 0.003 + 0.06

15. 4.62 + 0.078

16. 0.32 + 0.032 + 0.0032

17. 4.6 + 0.0005

18. 16.8 + 3.9

19. 1.62 + 2.078 + 3.1

20. 7.34 + 6 + 14.034

21. Add 0.68 to 1.7.

22. Find the sum of 3.28 and 14.021.

23. To 7.9 add 4 and 3.72.

24. Evaluate 7.9 + 0.62 + 5.

25. Find the sum of 8.6, 5 and 3.21.

SUBTRACTION OF DECIMALS

EXERCISE 5e Subtraction also may be done in the usual way, making sure that the decimal points are in line.

Find:

24.2 − 13.7	
	24.2
	− 13.7
24.2 − 13.7 = 10.5	10.5

1. 6.8 − 4.3	**5.** 0.0342 − 0.0021	**9.** 102.6 − 31.2
2. 9.6 − 1.8	**6.** 17.23 − 0.36	**10.** 7.32 − 0.67
3. 32.7 − 14.2	**7.** 3.273 − 1.032	**11.** 54.07 − 12.62
4. 0.62 − 0.21	**8.** 0.262 − 0.071	**12.** 7.063 − 0.124

It may be necessary to add noughts so that there is the same number of figures after the point in both cases.

4.623 − 1.7	
	4.623
	− 1.700
4.623 − 1.7 = 2.923	2.923

4.63 − 1.0342	
	4.6300
	− 1.0342
4.63 − 1.0342 = 3.5958	3.5958

2 − 1.4	
	2.0
	− 1.4
2 − 1.4 = 0.6	0.6

13. $3.26 - 0.2$

14. $3.2 - 0.26$

15. $14.23 - 11.1$

16. $6.8 - 4.14$

17. $11 - 8.6$

18. $7.98 - 0.098$

19. $7.098 - 0.98$

20. $3.2 - 0.428$

21. $11.2 - 0.0026$

22. $0.000\,32 - 0.000\,123$

23. $0.0073 - 0.0006$

24. $0.0073 - 0.006$

25. $0.006 - 0.000\,73$

26. $0.06 - 0.000\,73$

27. $6 - 0.73$

28. $6 - 0.073$

29. $7.3 - 0.06$

30. $730 - 0.6$

31. $0.73 - 0.000\,06$

32. $0.73 - 0.6$

33. Take 19.2 from 76.8.

34. Subtract 1.9 from 10.2.

35. From 0.168 subtract 0.019.

36. Evaluate $7.62 - 0.81$.

EXERCISE 5f   Find the value of:

1. $8.62 + 1.7$

2. $8.62 - 1.7$

3. $3.8 - 0.82$

4. $0.08 + 0.32 + 6.2$

5. $5 - 0.6$

6. $100 + 0.28$

7. $100 - 0.28$

8. $0.26 + 0.026$

9. $0.26 - 0.026$

10. $78.42 - 0.8$

11. $38.2 + 1.68$

12. $38.2 - 1.68$

13. $0.84 + 2 + 200$

14. $16 + 1.6 + 0.16$

15. $1.4 - 0.81$

16. $0.02 - 0.013$

17. $0.062 + 0.32$

18. $6.83 - 0.19$

19. $17.2 + 20 + 1.62$

20. $9.2 + 13.21 - 14.6$

Find the perimeter of the triangle (the perimeter is the distance all round).

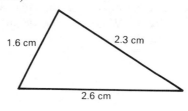

Perimeter = 1.6+2.3+2.6 cm
   = 6.5 cm

$$\begin{array}{r} 1.6 \\ 2.3 \\ +2.6 \\ \hline 6.5 \end{array}$$

21. Find the perimeter of the rectangle:

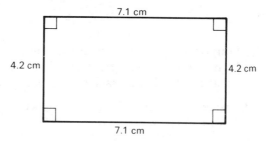

22. A piece of webbing is 7.6 m long. If 2.3 m is cut off, how much is left?

23. Find the total bill for three articles costing £5, £6.52 and £13.25.

24. The bill for two books came to £14.24. One book cost £3.72. What was the cost of the other one?

25. Add 2.32 and 0.68 and subtract the result from 4.

26. Find the perimeter of the quadrilateral:

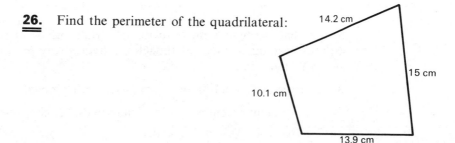

27. The bill for three meals was £6. The first meal cost £1.43 and the second £2.72. What was the cost of the third?

28. The perimeter of the quadrilateral is 19 cm. What is the length of the fourth side?

4.2 cm

3.1 cm

5.8 cm

MULTIPLICATION BY 10, 100, 1000, . . .

Consider $32 \times 10 = 320$. Writing 32 and 320 in headed columns gives

hundreds	tens	units
	3	2
3	2	0

Multiplying by 10 has made the number of units become the number of tens, and the number of tens has become the number of hundreds, so that all the figures have moved one place to the left.

Consider 0.2×10. When multiplied by 10, tenths become units $(\frac{1}{10} \times 10 = 1)$, so

units	tenths			units
0 .	2	$\times$ 10	=	2

Again the figure has moved one place to the left.

Multiplying by 100 means multiplying by 10 and then by 10 again, so the figures move 2 places to the left.

tens	units	tenths	hundredths	thousandths
	0 .	4	2	6 $\times$ 100
=	4	2 .	6	

Notice that the figures move to the left while the point stays put but without headings it looks as though the figures stay put and the point moves to the right.

When necessary we fill in an empty space with a nought.

units	tenths		hundreds	tens	units
4 .	2	$\times$ 100 =	4	2	0

EXERCISE 5g Find the value of:

> a) 368×100 $\qquad$ $368 \times 100 = 36\,800$
>
> b) 3.68×10 $\qquad$ $3.68 \times 10 = 36.8$
>
> c) 3.68×1000 $\qquad$ $3.68 \times 1000 = 3680$

1. 72×1000	**5.** 32.78×100	**9.** 72.81×1000
2. 8.24×10	**6.** $0.043 \times 10\,000$	**10.** $0.000\,006\,3 \times 10$
3. 0.0024×100	**7.** 0.0602×100	**11.** $0.007\,03 \times 100$
4. 46×10	**8.** 3.206×10	**12.** $0.0374 \times 10\,000$

DIVISION BY 10, 100, 1000, ...

When we divide by 10, hundreds become tens and tens become units.

hundreds	tens	units		tens	units
6	4	0	$\div 10 =$	6	4

The figures move one place to the right and the number becomes smaller but it looks as though the decimal point moves to the left so

$$2.72 \div 10 = 0.272$$

To divide by 100 the point is moved two places to the left.

To divide by 1000 the point is moved three places to the left.

EXERCISE 5h Find the value of:

> a) $3.2 \div 10$ $\qquad$ $3.2 \div 10 = 0.32$
>
> b) $320 \div 10\,000$ $\qquad$ $320 \div 10\,000 = 0.032$

1. $277.2 \div 100$	**5.** $27 \div 10$	**9.** $426 \div 10\,000$
2. $76.26 \div 10$	**6.** $6.8 \div 100$	**10.** $13.4 \div 10$
3. $0.000\,24 \div 10$	**7.** $0.26 \div 10$	**11.** $3.74 \div 1000$
4. $1.4 \div 100$	**8.** $15.8 \div 1000$	**12.** $0.92 \div 100$

MIXED MULTIPLICATION AND DIVISION

EXERCISE 5i   Find:

1. $1.6 \div 10$	**5.** 14.2×100	**9.** $140 \div 1000$
2. 1.6×10	**6.** 0.068×100	**10.** $7.8 \times 10\,000$
3. 0.078×100	**7.** $1.63 \div 100$	**11.** $24 \div 100$
4. $0.078 \div 100$	**8.** $2 \div 1000$	**12.** 0.063×1000

13. 0.32×10	**17.** 11.1×1000	**21.** $0.024 \div 100$
14. $7.9 \div 100$	**18.** $0.038 \div 100$	**22.** $0.3 \div 100\,000$
15. $0.000\,78 \times 100$	**19.** $0.38 \div 1000$	**23.** 0.0041×1000
16. $2.4 \div 10$	**20.** $3.8 \times 100\,000$	**24.** 0.1004×100

25. Share 42 m of string equally amongst 10 people.

26. Find the total cost of 100 articles at £1.52 each.

27. Evaluate $13.8 \div 100$ and 13.8×100.

28. Multiply 1.6 by 100 and then divide the result by 1000.

29. Add 16.2 and 1.26 and divide the result by 100.

30. Take 9.6 from 13.4 and divide the result by 1000.

DIVISION BY WHOLE NUMBERS

We can see that

$$\text{units}\quad\text{tenths}\qquad\qquad\text{units}\quad\text{tenths}$$
$$0\ .\ 6\quad \div 2 = \quad 0\ .\ 3$$

because 6 tenths $\div 2 = 3$ tenths. So we may divide by a whole number in the usual way as long as we keep the figures in the correct columns and the points are in line.

EXERCISE 5j

Find the value of $6.8 \div 2$

$$6.8 \div 2 = 3.4 \qquad \begin{array}{r} 2\overline{)6.8} \\ 3.4 \end{array} \qquad \text{(Keep the figures and points in line)}$$

Find the value of:

1. $0.4 \div 2$	**5.** $0.9 \div 9$	**9.** $42.6 \div 2$
2. $3.2 \div 2$	**6.** $0.95 \div 5$	**10.** $7.53 \div 3$
3. $0.63 \div 3$	**7.** $0.672 \div 3$	**11.** $6.56 \div 4$
4. $7.8 \div 3$	**8.** $26.6 \div 7$	**12.** $0.75 \div 5$

It may sometimes be necessary to fill spaces with noughts.

$0.00036 \div 3$

$0.00036 \div 3 = 0.00012$

$$3 \overline{)\, 0.00036}$$
$$0.00012$$

$0.45 \div 5$

$0.45 \div 5 = 0.09$

$$5 \overline{)\, 0.45}$$
$$0.09$$

$6.12 \div 3$

$6.12 \div 3 = 2.04$

$$3 \overline{)\, 6.12}$$
$$2.04$$

13. $0.057 \div 3$	**17.** $0.012 \div 6$	**21.** $1.232 \div 4$
14. $0.00065 \div 5$	**18.** $0.00036 \div 6$	**22.** $0.6552 \div 6$
15. $0.00872 \div 4$	**19.** $1.62 \div 2$	**23.** $0.0285 \div 5$
16. $0.168 \div 4$	**20.** $4.24 \div 4$	**24.** $0.1359 \div 3$

25. $0.0076 \div 4$	**29.** $6.3 \div 7$	**33.** $14.749 \div 7$
26. $0.81 \div 9$	**30.** $0.0636 \div 6$	**34.** $1.86 \div 3$
27. $0.5215 \div 5$	**31.** $0.038 \div 2$	**35.** $0.222 \div 6$
28. $0.000075 \div 5$	**32.** $4.62 \div 6$	**36.** $6.24 \div 8$

It may be necessary to add noughts at the end of a number in order to finish the division.

$2 \div 5$

$2 \div 5 = 0.4$

$$5 \overline{)\, 2.0}$$
$$0.4$$

$2.9 \div 8$

$2.9 \div 8 = 0.3625$

$$8 \overline{) 2.\overset{5}{9}\overset{2}{0}\overset{4}{0}00}$$
$$0.3625$$

Find the value of:

37. $6 \div 5$ **41.** $3.6 \div 5$ **45.** $9.1 \div 2$

38. $7.4 \div 4$ **42.** $0.0002 \div 5$ **46.** $0.00031 \div 2$

39. $0.83 \div 2$ **43.** $7.1 \div 8$ **47.** $9.4 \div 4$

40. $0.9 \div 6$ **44.** $7 \div 4$ **48.** $0.062 \div 5$

49. $0.5 \div 4$ **53.** $13 \div 5$ **57.** $6.83 \div 8$

50. $0.31 \div 8$ **54.** $0.3 \div 6$ **58.** $14.7 \div 6$

51. $2.6 \div 5$ **55.** $0.01 \div 4$ **59.** $2.3 \div 4$

52. $7.62 \div 4$ **56.** $3.014 \div 5$ **60.** $0.446 \div 8$

If we divide 7.8 m of tape equally amongst 5 people, how long a piece will they each have?

Length of each piece $= 7.8 \div 5$ m

$\qquad\qquad\qquad\qquad = 1.56$ m

$$5 \overline{) 7.\overset{2}{8}\overset{3}{0}}$$
$$1.56$$

61. The perimeter of a square is 14.6 cm. What is the length of a side?

62. Divide 32.6 m into 8 equal parts.

63. Share 14.3 kg equally between 2 people.

64. The perimeter of a regular pentagon (a five-sided figure with all the sides equal) is 16 cm. What is the length of one side?

65. Share £36 equally amongst 8 people.

LONG DIVISION

We can also use long division. The decimal point is used only in the original number and the answer, and not in the lines of working below these.

EXERCISE 5k Find the value of:

$2.56 \div 16$

$2.56 \div 16 = 0.16$

$$\begin{array}{r} 0.16 \\ 16\overline{)2.56} \\ \underline{16} \\ 96 \\ \underline{96} \end{array}$$

$4.2 \div 25$

$4.2 \div 25 = 0.168$

$$\begin{array}{r} 0.168 \\ 25\overline{)4.200} \\ \underline{25} \\ 170 \\ 150 \\ 200 \\ 200 \end{array}$$

1. $26.4 \div 24$
2. $2.1 \div 14$
3. $1.56 \div 13$
4. $9.45 \div 21$
5. $11.22 \div 22$
6. $80 \div 25$
7. $0.0615 \div 15$
8. $0.864 \div 24$
9. $8.48 \div 16$
10. $5.2 \div 20$
11. $7.84 \div 14$
12. $25.2 \div 36$

13. $35.52 \div 111$
14. $7.28 \div 28$
15. $1.296 \div 54$
16. $0.008\,05 \div 35$
17. $54.4 \div 17$
18. $21.93 \div 51$
19. $20.79 \div 99$
20. $0.014\,26 \div 20$
21. $23.4 \div 45$
22. $71.76 \div 23$
23. $39.48 \div 47$
24. $0.2556 \div 45$

CHANGING FRACTIONS TO DECIMALS (EXACT VALUES)

We may think of $\frac{3}{4}$ as $3 \div 4$ and hence write it as a decimal.

EXERCISE 5l

Express $\frac{3}{4}$ as a decimal.

$\frac{3}{4} = 3 \div 4 = 0.75$

$$\begin{array}{r} 4\overline{)3.00} \\ 0.75 \end{array}$$

Express the following fractions as decimals:

$$1\frac{2}{25}$$

$$1\frac{2}{25} = 1 + 2 \div 25$$
$$= 1 + 0.08$$
$$= 1.08$$

$$\begin{array}{r} 0.08 \\ 25\overline{)2.00} \\ 200 \end{array}$$

1. $\frac{1}{4}$ **3.** $\frac{3}{5}$ **5.** $\frac{1}{25}$ **7.** $\frac{5}{8}$ **9.** $\frac{3}{25}$

2. $\frac{3}{8}$ **4.** $\frac{5}{16}$ **6.** $2\frac{4}{5}$ **8.** $\frac{7}{16}$ **10.** $\frac{1}{32}$

STANDARD DECIMALS AND FRACTIONS

It is worth while knowing a few equivalent fractions and decimals. For example

$$\frac{1}{2} = 0.5 \qquad \frac{1}{4} = 0.25 \qquad \frac{1}{8} = 0.125$$

EXERCISE 5m Write the following decimals as fractions in their lowest terms, without any working if possible:

(Notice that $\quad \frac{2}{5} = \frac{4}{10} = 0.4$.)

1. 0.2   **3.** 0.8   **5.** 0.6   **7.** 0.9

2. 0.3   **4.** 0.75   **6.** 0.7   **8.** 0.05

Write down the following fractions as decimals:

9. $\frac{9}{10}$ **11.** $\frac{4}{5}$ **13.** $\frac{3}{100}$ **15.** $\frac{5}{8}$

10. $\frac{1}{4}$ **12.** $\frac{3}{8}$ **14.** $\frac{3}{4}$ **16.** $\frac{7}{100}$

MIXED EXERCISES

EXERCISE 5n   **1.** Write 0.02 as a fraction in its lowest terms.

       2. Write $\frac{9}{1000}$ and $\frac{91}{1000}$ as decimals.

       3. Add together 4.27, 31 and 1.6.

4. Subtract 1.82 from 4.2.

5. Divide 0.082 by 4.

6. Multiply 0.0301 by 100.

7. Express $\frac{7}{8}$ as a decimal.

8. Find the perimeter of the quadrilateral:

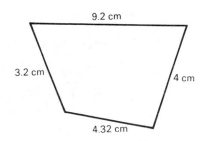

EXERCISE 5p **1.** Give 0.3 as a fraction.

2. Express $\frac{14}{100}$ as a decimal.

3. Find the sum of 16.2, 4.12 and 7.

4. Find the value of $0.062 \div 100$.

5. Divide 1.5 by 25.

6. Find the total bill for three books costing £4.26, £5 and £1.32.

7. Subtract 14.8 from 16.3.

8. Which is bigger, $\frac{2}{5}$ or 0.3?

EXERCISE 5q **1.** Give 0.008 as a fraction in its lowest terms.

2. Express $\frac{4}{5}$ as a decimal.

3. Add 14.2, 6, 0.38 and 7.21 together.

4. Subtract 14.96 from 100.

5. Divide 8.6 by 1000.

6. Evaluate $1.5 \div 6$.

7. Express $\frac{3}{16}$ as a decimal.

8. The perimeter of an equilateral triangle (all three sides are equal) is 14.4 cm. What is the length of one side?

EXERCISE 5r   **1.**   Express $\frac{1}{8}$ as a decimal.

2.   Find $8.2 - 1.92$.

3.   Divide 1.3 by 5.

4.   Add 4.2 and 0.28 and subtract 1.5 from the result.

5.   Express 0.09 as a fraction.

6.   Multiply 0.028 by ten thousand.

7.   Divide 42 by 15.

8.   I go into a shop with £6.22 and buy two articles, one costing £1.42 and the other £2.61. How much do I have left?

6 MULTIPLICATION AND DIVISION OF DECIMALS

LONG METHOD OF MULTIPLICATION

EXERCISE 6a   Calculate the following products:

$$0.3 \times 0.02$$

$$0.3 \times 0.02 = \frac{3}{10} \times \frac{2}{100}$$

$$= \frac{6}{1000}$$

$$= 0.006$$

$$0.06 \times 2$$

$$0.06 \times 2 = \frac{6}{100} \times \frac{2}{1}$$

$$= \frac{12}{100}$$

$$= 0.12$$

1. 0.04×0.2	**5.** 0.001×0.3	**9.** 0.08×0.01
2. 0.1×0.1	**6.** 0.4×0.0001	**10.** 0.0003×0.002
3. 0.003×6	**7.** 4×0.06	**11.** 0.9×0.02
4. 3×0.02	**8.** 0.4×0.0012	**12.** 0.004×2

SHORT METHOD OF MULTIPLICATION

In the examples above, if we add together the number of figures (including noughts) after the decimal points in the original two numbers, we get the number of figures after the point in the answer.

The number of figures after the point is called the number of decimal places. In the first example in Exercise 6a, 0.3 has one decimal place, 0.02 has two decimal places and the answer, 0.006, has three decimal places, which is the sum of one and two.

91

We can use this fact to work out 0.3×0.02 without using fractions. Multiply 3 by 2 ignoring the points; count up the number of decimal places after the points and then put the point in the correct position in the answer, writing in noughts where necessary, i.e. $0.3 \times 0.02 = 0.006$.

Any noughts that come after the point must be included when counting the decimal places.

EXERCISE 6b   Calculate the following products:

> 0.08×0.4
>
> $\underset{\text{(2 places)}}{0.08} \quad \times \quad \underset{\text{(1 place)}}{0.4} \quad = \quad \underset{\text{(3 places)}}{0.032} \qquad\qquad 8 \times 4 = 32$

> 6×0.002
>
> $\underset{\text{(0 places)}}{6} \quad \times \quad \underset{\text{(3 places)}}{0.002} \quad = \quad \underset{\text{(3 places)}}{0.012} \qquad\qquad 6 \times 2 = 12$

1.	0.6×0.3	**5.**	0.12×0.09	**9.**	0.08×0.08
2.	0.04×0.06	**6.**	0.07×0.0003	**10.**	3×0.0006
3.	0.009×2	**7.**	0.5×0.07	**11.**	0.7×0.06
4.	0.07×0.008	**8.**	8×0.6	**12.**	9×0.08

13.	0.07×12	**15.**	0.9×9	**17.**	7×0.011
14.	4×0.009	**16.**	0.0008×11	**18.**	0.04×7

Noughts appearing in the multiplication in the middle or at the right-hand end must also be considered when counting the places.

> 0.252×0.4
>
> $\underset{\text{(3 places)}}{0.252} \quad \times \quad \underset{\text{(1 place)}}{0.4} \quad = \quad \underset{\text{(4 places)}}{0.1008}$
>
> $\begin{array}{r} 252 \\ \times \quad 4 \\ \hline 1008 \end{array}$

> 2.5×6
>
> $\underset{\text{(1 place)}}{2.5} \quad \times \quad \underset{\text{(0 places)}}{6} \quad = \quad \underset{\text{(1 place)}}{15.0}$
>
> $\begin{array}{r} 25 \\ \times \quad 6 \\ \hline 150 \end{array}$

$$300 \times 0.2$$

$$\begin{array}{ccccc} 300 & \times & 0.2 & = & 60.0 \\ \text{(0 places)} & & \text{(1 place)} & & \text{(1 place)} \end{array} \qquad 300 \times 2 = 600$$

Calculate the following products:

19. 0.751×0.2	**23.** 400×0.6	**27.** 320×0.07
20. 3.2×0.5	**24.** 31.5×2	**28.** 0.4×0.0055
21. 0.35×4	**25.** 5.6×0.02	**29.** 0.5×0.06
22. 1.52×0.0006	**26.** 0.008×256	**30.** 0.04×0.352

31. 1.6×0.4	**35.** 4×1.6	**39.** 0.16×4
32. 1.6×0.5	**36.** 5×0.016	**40.** 0.0016×5
33. 160×0.004	**37.** $0.000\,04 \times 0.000\,16$	**41.** 0.072×0.6
34. 0.16×0.005	**38.** $16\,000 \times 0.05$	**42.** 310×0.04

MULTIPLICATION

EXERCISE 6c   Calculate the following products:

$$0.26 \times 1.3$$

$$\begin{array}{ccccc} 0.26 & \times & 1.3 & = & 0.338 \\ \text{(2 places)} & & \text{(1 place)} & & \text{(3 places)} \end{array}$$

$$\begin{array}{r} 26 \\ \times 13 \\ \hline 78 \\ 260 \\ \hline 338 \end{array}$$

1. 4.2×1.6	**5.** 310×1.4	**9.** 17.8×420
2. 52×0.24	**6.** 1.68×0.27	**10.** 3.2×37
3. 0.68×0.14	**7.** 13.2×2.5	**11.** 39×0.23
4. 48.2×26	**8.** 0.0082×0.034	**12.** 0.264×750

13. 14.4×4.5	**17.** 0.16×16	**21.** 14×0.123
14. 1.36×0.082	**18.** 0.0016×1600	**22.** 1.9×9.1
15. 0.081×0.032	**19.** 0.28×0.28	**23.** 8.2×2.8
16. 1.6×1.6	**20.** 0.34×0.31	**24.** 0.047×0.66

PROBLEMS

EXERCISE 6d

> Find the cost of 6 books at £2.35 each.
>
> Cost = £2.35 × 6   235
>
>   = £14.10   × 6
>   ____
>   1410

1. Find the cost of 10 articles at £32.50 each.

2. The perimeter of a square is 17.6 cm. Find the length of one side of the square.

3. Divide 26.6 kg into 7 equal parts.

4. Find the perimeter of a square of side 4.2 cm.

5. Find the cost of 62 notebooks at 68 p each, first in pence and then in pounds.

6. Multiply 3.2 by 0.6 and divide the result by 8.

7. If 68.25 m of ribbon is divided into 21 equal pieces, how long is each piece?

8. The length of a side of a regular twelve-sided polygon (a shape with 12 equal sides) is 4.2 m. Find the perimeter of the polygon.

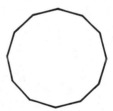

RECURRING DECIMALS

Consider the calculation

$$3 \div 4 = 0.75 \qquad 4 \overline{)\,3.00}$$
$$\qquad\qquad\qquad\qquad 0.75$$

By adding two noughts after the point we are able to finish the division and give an exact answer. Now consider

$$2 \div 3 = 0.666\ \ldots \qquad 3 \overline{)\,2.0000\ldots}$$
$$\qquad\qquad\qquad\qquad\qquad 0.6666\ldots$$

We can see that we will continue to obtain 6s for ever and we say that the 6 *recurs*.

Consider

$$31 \div 11 = 2.8181 \quad \ldots \qquad 11\,)\,\overline{31.0000\ldots}$$
$$2.8181\ldots$$

Here 81 recurs.

Sometimes it is one figure which is repeated and sometimes it is a group of figures. If one figure or a group continues to *recur* we have a *recurring decimal*.

EXERCISE 6e   Calculate:

> $0.2 \div 7$
>
> $0.2 \div 7 = 0.028\,571\,428\,571\ldots$
>
> $7\,)\,\overline{0.200\,000\,000\,000\ldots}$
> $0.028\,571\,428\,571\ldots$

1. $1.4 \div 6$ **3.** $4 \div 7$ **5.** $0.03 \div 7$

2. $0.03 \div 11$ **4.** $0.43 \div 3$ **6.** $1.1 \div 9$

Express the following fractions as decimals:

> $\dfrac{4}{3}$
>
> $\dfrac{4}{3} = 4 \div 3 = 1.333\ldots$
>
> $3\,)\,\overline{4.00\ldots}$
> $1.33\ldots$

7. $\dfrac{4}{9}$ **9.** $\dfrac{2}{11}$ **11.** $\dfrac{7}{9}$

8. $\dfrac{2}{3}$ **10.** $\dfrac{5}{7}$ **12.** $\dfrac{8}{7}$

To save writing so many figures we use a dot notation for recurring decimals.

For example

$$\frac{1}{6} = 1 \div 6 = 0.1666\ldots \qquad 6\,)\,\overline{1.000\ldots}$$
$$= 0.16\dot{6} \qquad\qquad 0.166\ldots$$

and

$$0.2 \div 7 = 0.0\dot{2}8\,571\,\dot{4} \qquad 7\,)\,\overline{0.2}$$
$$0.028\,571\,428\,571\,428$$

The dots are placed over the single recurring number or over the first and last figures of the recurring group.

13. Write the answers 1 to 12 above using the dot notation.

CORRECTING TO A GIVEN NUMBER OF DECIMAL PLACES

Often we need to know only the first few figures of a decimal. For instance, if we measure a length with an ordinary ruler we usually need an answer to the nearest $\frac{1}{10}$ cm and are not interested, or cannot see, how many $\frac{1}{100}$ cm are involved.

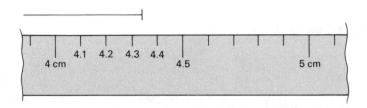

Look at this enlarged view of the end of a line which is being measured. We can see that with a more accurate measure we might be able to give the length as 4.34 cm. However on the given ruler we would probably measure it as 4.3 cm because we can see that the end of the line is nearer 4.3 than 4.4. We cannot give the exact length of the line but we can say that it is 4.3 cm long to the nearest $\frac{1}{10}$ cm. We write this as 4.3 cm correct to 1 decimal place.

Consider the numbers 0.62, 0.622, 0.625, 0.627 and 0.63. To compare them we write 0.62 as 0.620 and 0.63 as 0.630 so that each number has 3 figures after the point. When we write them in order in a column:

$$
\begin{array}{c}
0.620 \\
0.622 \\
0.625 \\
0.627 \\
0.630
\end{array}
$$

we can see that 0.622 is nearer to 0.620 than to 0.630 while 0.627 is nearer to 0.630 so we write

$0.62\!\mid\!2 = 0.62$ (correct to 2 decimal places)

$0.62\!\mid\!7 = 0.63$ (correct to 2 decimal places)

It is not so obvious what to do with 0.625 as it is halfway between 0.62 and 0.63. To save arguments, if the figure after the cut-off line is 5 or more we add 1 to the figure before the cut-off line, i.e. we round the number *up*, so we write

$0.62\!\mid\!5 = 0.63$ (correct to 2 decimal places)

EXERCISE 6f

Give 10.9315 correct to:

a) the nearest whole number   b) 1 decimal place
c) 3 decimal places.

a) 10.9315 = 11   (correct to the nearest whole number)

b) 10.9315 = 10.9   (correct to 1 decimal place)

c) 10.9315 = 10.932   (correct to 3 decimal places)

Give 4.699 and 0.007 correct to 2 decimal places.

4.699 = 4.70   (correct to 2 decimal places)

0.007 = 0.01   (correct to 2 decimal places)

Give the following numbers correct to 2 decimal places:

1.	0.328	**6.**	0.6947
2.	0.322	**7.**	0.8351
3.	1.2671	**8.**	3.927
4.	2.345	**9.**	0.0084
5.	0.0416	**10.**	3.9999

Give the following numbers correct to the nearest whole number:

11.	13.9	**16.**	6.783
12.	6.34	**17.**	109.7
13.	26.5	**18.**	6.145
14.	2.78	**19.**	74.09
15.	4.45	**20.**	3.9999

Give the following numbers correct to 3 decimal places:

21.	0.3627	**26.**	0.0843
22.	0.026 234	**27.**	0.084 47
23.	0.007 14	**28.**	0.3251
24.	0.0695	**29.**	0.032 51
25.	0.000 98	**30.**	3.9999

Give the following numbers correct to the number of decimal places indicated in the brackets:

31.	1.784	(1)		**36.**	1.639	(2)
32.	42.64	(1)		**37.**	1.639	(1)
33.	1.0092	(2)		**38.**	1.689	(nearest whole number)
34.	0.009 42	(4)		**39.**	3.4984	(2)
35.	0.7345	(3)		**40.**	3.4984	(1)

If we are asked to give an answer correct to a certain number of decimal places, we work out one more decimal place than is asked for. Then we can find the size of the last figure required.

EXERCISE 6g

Find $4.28 \div 6$ giving your answer correct to 2 decimal places.

$$4.28 \div 6 = 0.71\,3 \ldots$$
$$= 0.71 \quad \text{(correct to 2 decimal places)}$$

$$6\,)\,\overline{4.28\overset{2}{0}}$$
$$0.713\ldots$$

Calculate $302 \div 14$ correct to 1 decimal place.

$$302 \div 14 = 21.5\,7 \ldots$$
$$= 21.6 \quad \text{(correct to 1 decimal place)}$$

$$\begin{array}{r} 21.57\ldots \\ 14\,)\,\overline{302.00} \\ 28 \\ \hline 22 \\ 14 \\ \hline 80 \\ 70 \\ \hline 100 \\ 98 \end{array}$$

Calculate, giving your answers correct to 2 decimal places:

1.	$0.496 \div 3$		**5.**	$25.68 \div 9$	**9.**	$5.68 \div 24$
2.	$6.49 \div 7$		**6.**	$2.35 \div 15$	**10.**	$3.85 \div 101$
3.	$3.12 \div 9$		**7.**	$0.68 \div 16$	**11.**	$1.73 \div 8$
4.	$12.2 \div 6$		**8.**	$0.99 \div 21$	**12.**	$48.4 \div 51$

Calculate, giving your answers correct to 1 decimal place:

13.	$32.9 \div 8$	**17.**	$124 \div 17$	**21.**	$213 \div 22$
14.	$402 \div 7$	**18.**	$16.2 \div 14$	**22.**	$8.4 \div 13$
15.	$15.3 \div 6$	**19.**	$45 \div 21$	**23.**	$26 \div 15$
16.	$9.76 \div 11$	**20.**	$15.1 \div 16$	**24.**	$519 \div 19$

Find, giving your answers correct to 3 decimal places:

25.	$0.023 \div 4$	**29.**	$0.45 \div 12$	**33.**	$0.2584 \div 16$
26.	$0.123 \div 7$	**30.**	$0.012 \div 13$	**34.**	$0.321 \div 17$
27.	$1.25 \div 3$	**31.**	$0.654 \div 23$	**35.**	$1.26 \div 32$
28.	$0.23 \div 11$	**32.**	$0.98 \div 32$	**36.**	$0.88 \div 24$

CHANGING FRACTIONS TO DECIMALS

EXERCISE 6h

Give $\frac{4}{25}$ as a decimal.

$$\frac{4}{25} = 4 \div 25 = 0.16$$

(This is an exact answer.)

$$25 \overline{)\begin{array}{l} 0.16 \\ 4.00 \\ \underline{2\ 5} \\ 1\ 50 \end{array}}$$

Give $\frac{4}{7}$ as a decimal correct to 3 decimal places.

$$\frac{4}{7} = 4 \div 7 = 0.5714\ldots$$

$$= 0.571 \text{ (correct to 3 decimal places)}$$

(This is an approximate answer.)

$$7 \overline{)\begin{array}{l} 4.0000 \\ 0.5714\ldots \end{array}}$$

Give the following fractions as exact decimals:

1.	$\frac{5}{8}$	**2.**	$\frac{3}{40}$	**3.**	$\frac{3}{16}$	**4.**	$\frac{3}{5}$	**5.**	$\frac{9}{25}$
6.	$\frac{7}{50}$	**7.**	$\frac{1}{16}$	**8.**	$\frac{11}{8}$	**9.**	$\frac{13}{25}$	**10.**	$\frac{3}{80}$

Give the following fractions as decimals correct to 3 decimal places:

11. $\dfrac{3}{7}$ **13.** $\dfrac{1}{6}$ **15.** $\dfrac{9}{11}$ **17.** $\dfrac{8}{7}$ **19.** $\dfrac{1}{3}$

12. $\dfrac{4}{9}$ **14.** $\dfrac{2}{3}$ **16.** $\dfrac{6}{7}$ **18.** $\dfrac{1}{9}$ **20.** $\dfrac{4}{11}$

21. $\dfrac{3}{14}$ **23.** $\dfrac{6}{13}$ **25.** $\dfrac{3}{19}$ **27.** $\dfrac{4}{15}$ **29.** $\dfrac{3}{22}$

22. $\dfrac{4}{17}$ **24.** $\dfrac{4}{21}$ **26.** $\dfrac{3}{17}$ **28.** $\dfrac{7}{18}$ **30.** $\dfrac{4}{33}$

DIVISION BY DECIMALS

$0.012 \div 0.06$ can be written as $\frac{0.012}{0.06}$. We know how to divide by a whole number so we need to find an equivalent fraction with denominator 6 instead of 0.06. Now $0.06 \times 100 = 6$. Therefore we multiply the numerator and denominator by 100.

$$\frac{0.012}{0.06} = \frac{0.012 \times 100}{0.06 \ \times 100}$$

$$= \frac{1.2}{6}$$

$$= 0.2$$

To divide by a decimal, the denominator must be made into a whole number but the numerator need not be. We can write, for short,

$$0.012 \div 0.06 = \frac{0.01|2}{0.06|} = \frac{1.2}{6} \qquad \text{(keeping the points in line)}$$

the dotted line indicating where we want the point to be so as to make the denominator a whole number.

EXERCISE 6i   Find the exact answers to the following questions:

$0.024 \div 0.6$

$$0.024 \div 0.6 = \frac{0.0|24}{0.6|} = \frac{0.24}{6}$$

$$= 0.04$$

$$6 \overline{)\,0.24} \\ \,0.04$$

$64 \div 0.08$

$$64 \div 0.08 = \frac{64.00}{0.08} = \frac{6400}{8}$$

$$= 800$$

$$\begin{array}{r} 800 \\ 8 \overline{)6400} \end{array}$$

1.	$0.04 \div 0.2$	**5.**	$90 \div 0.02$	**9.**	$3.6 \div 0.06$
2.	$0.0006 \div 0.03$	**6.**	$0.48 \div 0.04$	**10.**	$3 \div 0.6$
3.	$4 \div 0.5$	**7.**	$0.032 \div 0.2$	**11.**	$6.5 \div 0.5$
4.	$0.8 \div 0.04$	**8.**	$3.6 \div 0.6$	**12.**	$8.4 \div 0.07$

13.	$72 \div 0.09$	**17.**	$0.9 \div 0.009$	**21.**	$0.000\,068\,4 \div 0.04$
14.	$1.08 \div 0.003$	**18.**	$0.92 \div 0.4$	**22.**	$20.8 \div 0.0004$
15.	$0.0012 \div 0.1$	**19.**	$16.8 \div 0.8$	**23.**	$0.0012 \div 0.3$
16.	$0.009 \div 0.9$	**20.**	$0.001\,32 \div 0.11$	**24.**	$4.8 \div 0.08$

$0.256 \div 1.6$

$$0.256 \div 1.6 = \frac{0.256}{1.6} = \frac{2.56}{16}$$

$$= 0.16$$

$$\begin{array}{r} 0.16 \\ 16 \overline{)2.56} \\ \underline{16} \\ 96 \\ 96 \end{array}$$

25.	$1.76 \div 2.2$	**29.**	$34.3 \div 1.4$	**33.**	$9.8 \div 1.4$
26.	$144 \div 0.16$	**30.**	$10.24 \div 3.2$	**34.**	$0.168 \div 0.14$
27.	$0.496 \div 1.6$	**31.**	$0.0204 \div 0.017$	**35.**	$1.35 \div 0.15$
28.	$0.0288 \div 0.18$	**32.**	$102.5 \div 2.5$	**36.**	$0.192 \div 2.4$

EXERCISE 6j

Find the value of $16.9 \div 0.3$ giving your answer correct to 1 decimal place.

$$16.9 \div 0.3 = \frac{16.9}{0.3} = \frac{169}{3}$$

$$= 56.33\ldots$$

$$= 56.3 \quad \text{(correct to 1 decimal place)}$$

$$\begin{array}{r} 3 \overline{)169.00} \\ 56.333\ldots \end{array}$$

Calculate, giving your answers correct to 2 decimal places:

1. $3.8 \div 0.6$
2. $0.59 \div 0.07$
3. $15 \div 0.9$
4. $5.633 \div 0.2$
5. $0.796 \div 1.1$
6. $1.25 \div 0.03$
7. $0.0024 \div 0.09$
8. $0.65 \div 0.7$
9. $0.0072 \div 0.007$
10. $5 \div 7$

Calculate, giving your answers correct to the number of decimal places indicated in the brackets:

11. $0.123 \div 6$ (2)
12. $2.3 \div 0.8$ (1)
13. $90 \div 11$ (1)
14. $0.0078 \div 0.09$ (3)
15. $12 \div 9$ (4)
16. $0.23 \div 0.007$ (1)
17. $16.2 \div 0.8$ (1)
18. $0.21 \div 6.5$ (3)
19. $85 \div 0.3$ (3)
20. $1.37 \div 0.8$ (1)

21. $56.9 \div 1.6$
    (nearest whole number)
22. $0.89 \div 0.23$ (1)
23. $0.75 \div 4.5$ (3)
24. $0.023 \div 0.021$ (1)
25. $3.2 \div 1.4$ (1)
26. $0.045 \div 0.012$
    (nearest whole number)
27. $12.3 \div 17$ (2)
28. $0.0054 \div 1.3$ (4)
29. $0.012 \div 0.021$ (2)
30. $0.52 \div 0.21$ (1)

MIXED MULTIPLICATION AND DIVISION

EXERCISE 6k   Calculate, giving your answers exactly:

1. 0.48×0.3
2. $0.48 \div 0.3$
3. 2.56×0.02
4. $2.56 \div 0.02$
5. 3.6×0.8
6. 9.6×0.6
7. 0.0042×0.03
8. $0.0042 \div 0.03$
9. 16.8×0.4
10. $1.68 \div 0.4$
11. 20.4×0.6
12. $5.04 \div 0.06$

$$\frac{0.12 \times 3}{0.006}$$

$$\frac{0.12 \times 3}{0.006} = \frac{0.36}{0.006} \qquad\qquad 12 \times 3 = 36$$

$$= \frac{360}{6}$$

$$= 60$$

Find the value of:

13. $\dfrac{0.2 \times 0.6}{0.4}$ 　　　**16.** $\dfrac{3.2}{4 \times 0.2}$ 　　　**19.** $\dfrac{2.5 \times 0.7}{3.5 \times 4}$

14. $\dfrac{1.2 \times 0.04}{0.3}$ 　　　**17.** $\dfrac{3}{0.6 \times 0.5}$ 　　　**20.** $\dfrac{5.6 \times 0.8}{6.4}$

15. $\dfrac{4.8 \times 0.2}{0.6 \times 0.4}$ 　　**18.** $\dfrac{4.4 \times 0.3}{11}$ 　　　**21.** $\dfrac{0.9 \times 4}{0.5 \times 0.6}$

RELATIVE SIZES

To compare the sizes of numbers they need to be in the same form, either as fractions with the same denominators, or as decimals.

EXERCISE 6I

Express 0.82, $\frac{4}{5}$, $\frac{9}{11}$ as decimals where necessary and write them in order of size with the smallest first.

$$\frac{4}{5} = 0.8$$

$$\frac{9}{11} = 0.8181\ldots \qquad\qquad 11 \overline{)9.000}$$
$$\qquad\qquad\qquad\qquad\qquad\qquad 0.8181\ldots$$

In order of size:　　$\dfrac{4}{5}$, $\dfrac{9}{11}$, 0.82

Express the following sets of numbers as decimals or as fractions and write them in order of size with the smallest first:

1. $\dfrac{1}{4}$, 0.2

2. $\dfrac{2}{5}$, $\dfrac{4}{9}$

3. $\dfrac{1}{2}$, $\dfrac{4}{9}$

4. $\dfrac{1}{3}$, 0.3, $\dfrac{3}{11}$

5. $\dfrac{8}{9}$, 0.9, $\dfrac{7}{8}$

6. $\dfrac{3}{4}$, $\dfrac{17}{20}$

7. $\dfrac{3}{8}$, $\dfrac{9}{25}$, 0.35

8. $\dfrac{3}{5}$, $\dfrac{4}{7}$, 0.59

9. $\dfrac{3}{7}$, $\dfrac{5}{11}$, $\dfrac{6}{13}$

10. $0.\dot{7}$, $\dfrac{8}{11}$

11. $0.\dot{3}$, $\dfrac{5}{12}$

12. $\dfrac{1}{2}$, 0.45, $\dfrac{9}{19}$

MIXED EXERCISES

EXERCISE 6m **1.** Multiply 0.68 by
a) 10   b) 1000.

2. Express $\frac{7}{8}$ as a decimal.

3. Give the following numbers correct to 2 decimal places:
a) 3.126   b) 0.075   c) 2.999

4. Add 3.86, 14.2 and 2.078.

5. Multiply 3.2 by 1.4.

6. Divide 8.2 by 5.

7. Subtract 4.28 from 16.1.

8. Which is larger, 6.6 or $6\frac{2}{3}$? Why?

EXERCISE 6n **1.** Express 0.06 as a fraction in its lowest terms.

2. Divide 6.24 by
a) 100   b) 12.

3. Add 3.2 and 0.9 and subtract the result from 5.8.

4. The perimeter of an equilateral triangle (a triangle with three equal sides) is 19.2 cm. Find the length of one side.

5. Divide 0.0432 by 0.9.

6. Express $\frac{6}{25}$ as a decimal.

7. Find the cost of 24 articles at £2.32 each.

8. Give 7.7815 correct to

a) nearest whole number   b) 1 decimal place

c) 3 decimal places.

EXERCISE 6p **1.** Give $\frac{5}{7}$ as a recurring decimal.

2. Divide   a) 6.4   b) 0.064   by 100.

3. Multiply 14.8 by 1.1.

4. Express 0.62 as a fraction in its lowest terms.

5. Add 6.7, 0.67, 0.067 and 0.0067 together.

6. Divide 16.4 by 8.

7. Which is bigger, 0.7 or $\frac{7}{9}$?

8. How many pieces of ribbon of length 0.3 m can be cut from a piece 7.5 m long?

EXERCISE 6q **1.** Express $\frac{4}{25}$ as a decimal.

2. Find $6.43 \div 0.7$ correct to 3 decimal places.

3. Find 0.06×0.06.

4. Express 0.0095 as a fraction in its lowest terms.

5. Find $13.8 + 2.43 - 1.6$.

6. Find the cost of 3.5 m of ribbon at 17 p a metre.

7. Find $\dfrac{0.6 \times 0.3}{0.09}$.

8. Write $0.\dot{6}$ in another way as a decimal. Why is it not easy to find $0.7 - 0.\dot{6}$?

7 METRIC UNITS

Whenever we want to measure a length, or weigh an object, we find the length or weight in standard units. We might for instance give the length of a line in millimetres or the weight of a bag of apples in pounds. The millimetre belongs to a set of units called the metric system. The pound is one of the imperial units.

UNITS OF LENGTH

The basic unit of length is the *metre* (m). To get an idea of how long a metre is, remember that a standard bed is about 2 m long. However, a metre is not a useful unit for measuring either very large things or very small things so we need larger units and smaller units.

We get the larger unit by multiplying the metre by 1000. We get smaller units by dividing the metre into 100 parts or 1000 parts.

1000 metres is called 1 kilometre   (km)

(It takes about 15 minutes to walk a distance of 1 km.)

$$\frac{1}{100} \text{ of a metre is called 1 centimetre   (cm)}$$

$$\frac{1}{1000} \text{ of a metre is called 1 millimetre   (mm)}$$

(You can see centimetres and millimetres on your ruler.)

EXERCISE 7a **1.** Which metric unit would you use to measure

    a) the length of your classroom
    b) the length of your pencil
    c) the length of a soccer pitch
    d) the distance from Manchester to London
    e) the length of a page in this book
    f) the thickness of your exercise book?

2. Use your ruler to draw a line of length

a) 10 cm	e) 20 mm	i) 25 mm
b) 3 cm	f) 4 cm	j) 16 mm
c) 15 cm	g) 15 mm	k) 5 cm
d) 50 mm	h) 12 cm	l) 75 mm

3. Estimate the length, in centimetres, of the following lines:

a) ————————————

b) ————

c) ——————————————

d) ——

e) ————————————————————

Now use your ruler to measure each line.

4. Estimate the length, in millimetres, of the following lines:

a) ————————   c) —   e) ————————

b) ——   d) ——————

Now use your ruler to measure each line.

5. Use a straight edge (not a ruler with a scale) to draw a line that is approximately

a) 10 cm long   b) 5 cm long   c) 15 cm long   d) 20 mm long

Now measure each line to see how good your approximation was.

6. Estimate the width of your classroom in metres.

7. Estimate the length of your classroom in metres.

8. Measure the length and width of your exercise book in centimetres. Draw a rough sketch of your book with the measurements on it. Find the perimeter (the distance all round) of your book.

9. Each side of a square is 10 cm long. Draw a rough sketch of the square with the measurements on it. Calculate the perimeter of the square.

10. A sheet is 200 cm wide and 250 cm long. What is the perimeter of the sheet?

CHANGING FROM LARGE UNITS TO SMALLER UNITS

The metric units of length are the kilometre, the metre, the centimetre and the millimetre where

$$1 \text{ km} = 1000 \text{ m} \qquad\qquad 1 \text{ m} = 100 \text{ cm}$$

$$1 \text{ m} = 1000 \text{ mm} \qquad\qquad 1 \text{ cm} = 10 \text{ mm}$$

EXERCISE 7b Express the given quantity in terms of the unit in brackets:

3 km   (m)

$$3\text{ km} = 3 \times 1000\text{ m}$$
$$= 3000\text{ m}$$

3.5 m   (cm)

$$3.5\text{ m} = 3.5 \times 100\text{ cm}$$
$$= 350\text{ cm}$$

1.	2 m	(cm)	**5.**	12 km	(m)	**9.**	3 m	(mm)
2.	5 km	(m)	**6.**	15 cm	(mm)	**10.**	2 km	(mm)
3.	3 cm	(mm)	**7.**	6 m	(mm)	**11.**	5 m	(cm)
4.	4 m	(cm)	**8.**	1 km	(cm)	**12.**	7 m	(mm)
13.	1.5 m	(cm)	**17.**	1.9 m	(mm)	**21.**	3.8 cm	(mm)
14.	2.3 cm	(mm)	**18.**	3.5 km	(m)	**22.**	9.2 m	(mm)
15.	4.6 km	(m)	**19.**	2.7 m	(cm)	**23.**	2.3 km	(m)
16.	3.7 m	(mm)	**20.**	1.9 km	(cm)	**24.**	8.4 m	(cm)

UNITS OF WEIGHT

The most familiar units used for weighing are the kilogram (kg) and the gram (g).

Most groceries that are sold in tins or packets have weights given in grams. For example the weight of the most common packet of butter is 250 g. One eating apple weighs roughly 100 g, so the gram is a small unit of weight. Kilograms are used to give the weights of sugar and flour: the weight of the most common bag of sugar is 1 kg and the most common bag of flour weighs 1.5 kg.

For weighing large loads (coal or steel for example) a larger unit of weight is needed, and we use the tonne (t). For weighing very small quantities (for example the weight of a particular drug in one pill) we use the milligram (mg).

The relationships between these weights are

$$1\,t \quad = \quad 1000\,kg$$
$$1\,kg \quad = \quad 1000\,g$$
$$1\,g \quad = \quad 1000\,mg$$

EXERCISE 7c   Express each quantity in terms of the unit given in brackets:

> 2 t   (g)
>
> $$2\,t = 2 \times 1000\,kg$$
> $$= 2000\,kg$$
> $$= 2000 \times 1000\,g$$
> $$= 2\,000\,000\,g$$

1.	12 t	(kg)	**5.**	1 kg	(mg)	**9.**	4 kg	(g)
2.	3 kg	(g)	**6.**	13 kg	(g)	**10.**	2 kg	(mg)
3.	5 g	(mg)	**7.**	6 g	(mg)	**11.**	3 t	(kg)
4.	1 t	(g)	**8.**	2 t	(g)	**12.**	4 g	(mg)
13.	1.5 kg	(g)	**17.**	5.2 kg	(mg)	**21.**	7.3 g	(mg)
14.	2.7 t	(kg)	**18.**	0.6 g	(mg)	**22.**	0.3 kg	(mg)
15.	1.8 g	(mg)	**19.**	11.3 t	(kg)	**23.**	0.5 t	(kg)
16.	0.7 t	(kg)	**20.**	2.5 kg	(g)	**24.**	0.8 g	(mg)

MIXED UNITS

When you use your ruler to measure a line, you will probably find that the line is not an exact number of centimetres. For example the width of this page is 16 cm and 4 mm. We can say that the width of this page is 16 cm 4 mm or we could give the width in millimetres alone.

Now $$16\,cm = 16 \times 10\,mm$$
$$= 160\,mm$$

So $$16\,cm\ 4\,mm = 164\,mm$$

EXERCISE 7d   Express each quantity in terms of the units given in brackets:

> 4 kg 50 g   (g)
>
> $$4 \text{ kg} = 4 \times 1000 \text{ g}$$
> $$= 4000 \text{ g}$$
>
> Therefore 4 kg 50 g = 4050 g.

1.	1 m 36 cm	(cm)	**6.**	3 km 20 m	(m)
2.	3 cm 5 mm	(mm)	**7.**	5 m 2 cm	(cm)
3.	1 km 50 m	(m)	**8.**	5 km 500 m	(m)
4.	4 cm 8 mm	(mm)	**9.**	20 cm 2 mm	(mm)
5.	2 m 7 cm	(cm)	**10.**	8 m 9 mm	(mm)
11.	3 kg 500 g	(g)	**16.**	1 kg 20 g	(g)
12.	2 kg 8 g	(g)	**17.**	1 g 250 mg	(mg)
13.	5 g 500 mg	(mg)	**18.**	3 kg 550 g	(g)
14.	2 t 800 kg	(kg)	**19.**	2 t 50 kg	(kg)
15.	3 t 250 kg	(kg)	**20.**	1 kg 10 g	(g)

CHANGING FROM SMALL UNITS TO BIGGER UNITS

EXERCISE 7e   Express the first quantity in terms of the units given in brackets:

> 400 cm   (m)
>
> $$400 \text{ cm} = 400 \div 100 \text{ m}$$
> $$= 4 \text{ m}$$

> 580 g   (kg)
>
> $$580 \text{ g} = 580 \div 1000 \text{ kg}$$
> $$= 0.58 \text{ kg}$$

In questions 1 to 20, express the first quantity in terms of the units given in brackets:

1.	300 mm	(cm)		**6.**	72 m	(km)
2.	6000 m	(km)		**7.**	12 cm	(m)
3.	150 cm	(m)		**8.**	88 mm	(cm)
4.	250 mm	(cm)		**9.**	1250 mm	(m)
5.	1600 m	(km)		**10.**	2850 m	(km)
11.	1500 kg	(t)		**16.**	86 kg	(t)
12.	3680 g	(kg)		**17.**	560 g	(kg)
13.	1500 mg	(g)		**18.**	28 mg	(g)
14.	5020 g	(kg)		**19.**	190 kg	(t)
15.	3800 kg	(t)		**20.**	86 g	(kg)

In questions 21 to 40 express the given quantity in terms of the units given in brackets:

> 5 m 36 cm   (m)
>
> $$36\,cm = 36 \div 100\,m$$
> $$= 0.36\,m$$
> $$So\ \ 5\,m\ 36\,cm = 5.36\,m.$$

21.	3 m 45 cm	(m)		**26.**	5 m 3 cm	(m)
22.	8 cm 4 mm	(cm)		**27.**	7 km 5 m	(km)
23.	11 km 2 m	(km)		**28.**	4 m 5 mm	(m)
24.	2 km 42 m	(km)		**29.**	1 km 10 cm	(km)
25.	4 cm 4 mm	(cm)		**30.**	8 cm 5 mm	(km)
31.	5 kg 142 g	(kg)		**36.**	4 g 111 mg	(g)
32.	48 g 171 mg	(g)		**37.**	1 t 56 kg	(t)
33.	9 kg 8 g	(kg)		**38.**	5 g 3 mg	(g)
34.	9 g 88 mg	(g)		**39.**	250 g 500 mg	(kg)
35.	12 kg 19 g	(kg)		**40.**	850 kg 550 g	(t)

ADDING AND SUBTRACTING METRIC QUANTITIES

EXERCISE 7f   Quantities must be expressed in the same units before they are added or subtracted.

Find $1\,kg + 158\,g$ in a) grams b) kilograms.

a) $\qquad\qquad 1\,kg = 1000\,g$

$\qquad \therefore\ 1\,kg + 158\,g = 1158\,g$

($\therefore$ means "therefore" or "it follows that")

b) $\qquad\qquad 158\,g = 158 \div 1000\,kg$

$\qquad\qquad\qquad\quad = 0.158\,kg$

$\qquad \therefore\ 1\,kg + 158\,g = 1.158\,kg$

Find the sum of $5\,m$, $4\,cm$ and $97\,mm$ in
a) metres b) centimetres.

a) $\qquad\qquad 4\,cm = 4 \div 100\,m = 0.04\,m$

$\qquad\qquad 97\,mm = 97 \div 1000\,m = 0.097\,m$

$\therefore\ 5\,m + 4\,cm + 97\,mm = (5 + 0.04 + 0.097)\,m$

$\qquad\qquad\qquad\qquad\qquad = 5.137\,m$

b) $\qquad\qquad 5\,m = 5 \times 100\,cm = 500\,cm$

$\qquad\qquad 97\,mm = 97 \div 10\,cm = 9.7\,cm$

$\therefore\ 5\,m + 4\,cm + 97\,mm = (500 + 4 + 9.7)\,cm$

$\qquad\qquad\qquad\qquad\qquad = 513.7\,cm$

Find, giving your answer in metres:

1. $5\,m + 86\,cm$

2. $92\,cm + 115\,mm$

3. $3\,km + 136\,cm$

4. $51\,m + 3\,km$

5. $36\,cm + 87\,mm + 520\,cm$

6. $120\,mm + 53\,cm + 4\,m$

Find, giving your answer in millimetres:

7. $36\,cm + 80\,mm$

8. $5\,cm + 5\,mm$

9. $1\,m + 82\,cm$

10. $2\,m + 45\,cm + 6\,mm$

11. $3\,cm + 5\,m + 2.9\,cm$

12. $34\,cm + 18\,mm + 1\,m$

Find, giving your answer in grams:

13. 3 kg + 250 g

14. 5 kg + 115 g

15. 5.8 kg + 9.3 kg

16. 1 kg + 0.8 kg + 750 g

17. 116 g + 0.93 kg + 680 mg

18. 248 g + 0.06 kg + 730 mg

Find, expressing your answer in kilograms:

19. 2 t + 580 kg

20. 1.8 t + 562 kg

21. 390 g + 1.83 kg

22. 1.6 t + 3.9 kg + 2500 g

23. 1.03 t + 9.6 kg + 0.05 t

24. 5.4 t + 272 kg + 0.3 t

Find, expressing your answer in the units given in brackets:

25. 8 m − 52 cm   (cm)

26. 52 mm + 87 cm   (m)

27. 1.3 kg − 150 g   (g)

28. 1.3 m − 564 mm   (cm)

29. 2.05 t + 592 kg   (kg)

30. 20 g − 150 mg   (mg)

31. 36 kg − 580 g   (g)

32. 1.5 t − 590 kg   (kg)

33. 3.9 m + 582 mm   (cm)

34. 0.3 m − 29.5 cm   (mm)

MULTIPLYING METRIC UNITS

EXERCISE 7g   Calculate, expressing your answer in the units given in brackets:

3 × 2 g 741 mg   (g)

$$2 \text{ g } 741 \text{ mg} = 2.741 \text{ g}$$

$$\therefore 3 \times 2 \text{ g } 741 \text{ mg} = 3 \times 2.741 \text{ g}$$

$$= 8.223 \text{ g}$$

$$\begin{array}{r} 2741 \\ \times\ \ \ 3 \\ \hline 8223 \end{array}$$

1. 4 × 3 kg 385 g   (g)

2. 9 × 5 m 88 mm   (mm)

3. 3 × 4 kg 521 g   (kg)

4. 5 × 2 m 51 cm   (m)

5. 10 × 3 t 200 kg   (t)

6. 2 × 5 cm 3 mm   (cm)

7. 6 × 2 g 561 mg   (mg)

8. 8 × 3 km 56 m   (km)

9. 3 × 7 t 590 kg   (t)

10. 7 × 2 km 320 m   (m)

PROBLEMS

EXERCISE 7h

> Find, in kilograms, the total weight of a bag of flour weighing 1.5 kg, a jar of jam weighing 450 g and a packet of rice weighing 500 g.
>
> $$\text{The jar of jam weighs } 450\,\text{g} = 450 \div 1000\,\text{kg}$$
> $$= 0.45\,\text{kg}$$
>
> $$\text{The packet of rice weighs } 500\,\text{g} = 500 \div 1000\,\text{kg}$$
> $$= 0.5\,\text{kg}$$
>
> $$\text{The total weight} = (1.5 + 0.45 + 0.5)\,\text{kg}$$
> $$= 2.45\,\text{kg}$$

1. Find the sum, in metres, of 5 m, 52 cm, 420 cm.

2. Find the sum, in grams, of 1 kg, 260 g, 580 g.

3. Subtract 52 kg from 0.8 t, giving your answer in kilograms.

4. Find the difference, in grams, between 5 g and 890 mg.

5. Find the total length, in millimetres, of a piece of wood 82 cm long and another piece of wood 260 mm long.

6. Find the total weight, in kilograms, of 500 g of butter, 2 kg of potatoes, 1.5 kg of flour.

7. One tin of baked beans weighs 220 g. What is the weight, in kilograms, of ten of these tins?

8. One fence post is 150 cm long. What length of wood, in metres, is needed to make ten such fence posts?

9. Find the perimeter of a square if each side is of length 8.3 cm. Give your answer in centimetres.

10. A wooden vegetable crate and its contents weigh 6.5 kg. If the crate weighs 1.2 kg what is the weight of its contents?

MONEY UNITS

Many countries use units of money that are divided into hundredths. For example

UK	1 pound (£) = 100 pence (p)
USA	1 dollar ($) = 100 cents (c)
France	1 franc = 100 centimes
Germany	1 mark = 100 pfennigs

EXERCISE 7i   Express each quantity in terms of the units given in brackets:

1. 7 dollars   (cents)

2. £6   (pence)

3. 8 marks   (pfennigs)

4. 13 francs   (centimes)

5. 7 francs 35 centimes   (centimes)

6. 43 dollars 81 cents   (cents)

7. 11 marks 3 pfennigs   (pfennigs)

8. £6 15 p   (pence)

9. £2 10 p   (pence)

10. £5 4 p   (pence)

---

420 p   (£)

$$420\,p = £4.20$$

(Note that we always give pounds to 2 decimal places so we write £4.20 rather than £4.2. Other currencies are written in the same way.)

---

11. 126 p   (£)

12. 350 cents   (dollars)

13. 190 p   (£)

14. 350 pfennigs   (marks)

15. 43 dollars 7 cents   (dollars)

16. 228 p   (£)

17. 3 marks 47 pfennigs   (marks)

18. 580 p   (£)

19. 11 francs 9 centimes   (francs)

20. £6 8 p   (£)

21. One tin of baked beans costs 32 p. Find the cost, in pounds, of ten of these tins.

22. Find the total cost, in dollars, of a book costing $4, a pencil costing 30 cents and a magazine costing 75 cents.

23. Find the cost, in pounds, of 20 litres of petrol at 48 p a litre.

24. One can of cola costs 50 pfennigs. Find the cost, in marks, of twelve such cans.

25. Find the cost of ten stamps at $16\frac{1}{2}$ p each. If you paid for these stamps with a £5 note, how much change would you get?

MIXED PROBLEMS

EXERCISE 7j

A girl takes to school a bag with books in it, a shoe bag and a clarinet. The bag and its contents weigh 2.5 kg, the shoe bag weighs 900 g and the clarinet weighs 1 kg 900 g. What is the total weight, in kg, that the girl carries?

$$2.5 \text{ kg}$$
$$900 \text{ g} = 0.9 \text{ kg}$$
$$1 \text{ kg } 900 \text{ g} = \underline{1.9 \text{ kg}}$$
$$\text{Total weight} = 5.3 \text{ kg}$$

1. A rectangular sheet of paper measures 32 cm by 17 cm. What is its perimeter
a) in centimetres   b) in millimetres?

2. A girl travels to school by walking 450 m to the bus stop and then travelling 1 km 650 m by bus. The distance she walks after getting off the bus is 130 m. What distance is her total journey in kilometres?

3. A man takes three parcels to the Post Office and has them weighed. One parcel weighs 4 kg 37 g, the weight of the second is 3 kg 982 g and the third one weighs 1 kg 173 g. What is their total weight in kilograms?

4. A rectangular field is 947 m long and 581 m wide. What is the perimeter of the field? How many metres of fencing would be needed to go round the field leaving space for two gates each 3 m wide?

5. Wood is sometimes sold by the 'metric foot'. A metric foot is 30 cm. A man buys a length of wood which is 12 metric feet long. How long is the piece of wood in metres?

6. A freight train has five trucks. Two of them are carrying 15 t 880 kg each. Another has a load of 14 t 700 kg and the last two are each loaded with 24 t 600 kg. What is the total weight, in tonnes, of the contents of the five trucks? If the weight of each truck is 5 t 260 kg, what is the combined weight of the trucks and their contents?

7. A boy delivers newspapers by bicycle. The weight of the bicycle is 15.8 kg and the boy weighs 51.3 kg. At the beginning of the round the newspapers weigh 9.8 kg. What is the total weight of the boy and his bicycle loaded with newspapers? What is the weight when he has delivered half the newspapers?

8. Newtown is 5.62 km from Old Town and Old Town is 3.87 km from Castletown. If a car goes from Newtown to Old Town, then to Castletown and finally back to Old Town, how many metres has it travelled? At the beginning of its journey the car had enough petrol to go 27 km. At the end of its journey how much further could it go before running out of petrol?

9. In France three letters are posted to various foreign countries. The stamps cost 1 franc 50 centimes, 2 francs 80 centimes and 1 franc 90 centimes. What is the total cost of the stamps? How much change would be given if the stamps were bought with a 10 franc piece

a) in francs   b) in centimes?

10. A man takes a parcel to the Post Office. It weighs 3 kg 750 g. The cost of postage is 45 p per 250 g. How much does the parcel cost to post?

MIXED EXERCISES

EXERCISE 7k   Express the given quantity in terms of the units given in brackets:

1. 4 km (m)		**4.** 250 g (kg)		**7.** 1 m 50 cm (m)	
2. 30 g (kg)		**5.** 0.03 km (cm)		**8.** 2.8 cm (mm)	
3. 3.5 m (cm)		**6.** 1250 m (km)		**9.** 65 g (kg)	

10. A tin of meat weighs 429 g. What is the weight, in kilograms, of ten such tins?

EXERCISE 7l   Express the given quantity in terms of the units in brackets:

1. 236 cm   (m)
2. 0.02 m   (mm)
3. 5 kg   (g)
4. 500 mg   (g)

5. 4 km 250 m   (km)
6. 3.6 t   (kg)
7. 2 kg 350 g   (kg)
8. 2 g   (mg)

9. Each side of a square is 65 cm long. What is the perimeter of the square, in metres?

EXERCISE 7m   Express the given quantity in terms of the units in brackets:

1. 5.78 t   (kg)
2. £3 54 p   (p)
3. 350 kg   (t)
4. 0.155 mm   (cm)

5. 1 t 560 kg   (t)
6. 780 centimes   (francs)
7. 0.36 g   (mg)
8. 2 km 50 m   (km)

9. A bus weighs 5 t 430 kg and carries 44 passengers each of whom is assumed to weigh 72 kg. Find the weight, in tonnes, of the bus and passengers when it is fully loaded.

EXERCISE 7n   Express the given quantity in terms of the units in brackets:

1. 4 cm 2 mm   (cm)
2. 350 g   (kg)
3. £1 52 p   (£)
4. 283 m   (km)

5. 36 mm   (cm)
6. 0.47 m   (mm)
7. 36 cm   (m)
8. 1356 mg   (g)

9. A bag containing 2 p coins weighs 2.492 kg. If the weight of one coin is 7.12 g find the value in pounds of the coins in the bag.

8 IMPERIAL UNITS

UNITS OF LENGTH

Some imperial units are still used. For instance distances on road signs are still given in miles. One mile is roughly equivalent to $1\frac{1}{2}$ km. A better approximation is

5 miles is about 8 kilometres

Yards, feet and inches are other imperial units of length that are still used. In this system units are not always divided into ten parts to give smaller units so we have to learn "tables".

$$12 \text{ inches (in)} = 1 \text{ foot (ft)}$$

$$3 \text{ feet} = 1 \text{ yard (yd)}$$

$$1760 \text{ yards} = 1 \text{ mile}$$

EXERCISE 8a Express the given quantity in the units in brackets:

> 2 ft 5 in (in)
>
> $$2\,\text{ft} = 2 \times 12\,\text{in}$$
> $$= 24\,\text{in}$$
> $$\therefore \ 2\,\text{ft}\ 5\,\text{in} = 24 + 5\,\text{in}$$
> $$= 29\,\text{in}$$

1. 5 ft 8 in   (in)   **6.** 2 miles 800 yd   (yd)

2. 4 yd 2 ft   (ft)   **7.** 5 yd 2 ft   (ft)

3. 1 mile 49 yd (yd)   **8.** 10 ft 3 in   (in)

4. 2 ft 11 in   (in)   **9.** 9 yd 1 ft   (ft)

5. 8 ft 4 in   (in)   **10.** 9 ft 10 in   (in)

> 52 in (ft and in)
>
> $$52\,\text{in} = 52 \div 12\,\text{ft}$$
> $$= 4\,\text{ft}\ 4\,\text{in}$$
>
> $$\begin{array}{r} 4 \quad \text{r}\,4 \\ 12\overline{)\,52} \end{array}$$

11.	36 in	(ft)		

11. 36 in   (ft)   **16.** 2000 yd   (miles and yd)

12. 29 in   (ft and in)   **17.** 75 in   (ft and in)

13. 86 in   (ft and in)   **18.** 100 ft   (yd and ft)

14. 9 ft   (yd)   **19.** 120 in   (ft and in)

15. 13 ft   (yd and ft)   **20.** 30 000 yd   (miles and yd)

UNITS OF WEIGHT

The imperial units of weight that are still used are pounds and ounces. Other units of weight that you may still see are hundredweights and tons (not to be confused with tonnes).

$$16 \text{ ounces (oz)} = 1 \text{ pound (lb)}$$

$$112 \text{ pounds} = 1 \text{ hundredweight (cwt)}$$

$$20 \text{ hundredweight} = 1 \text{ ton}$$

EXERCISE 8b   Express the given quantity in terms of the units given in brackets:

1. 2 lb 6 oz   (oz)   **6.** 24 oz   (lb and oz)

2. 1 lb 12 oz   (oz)   **7.** 18 oz   (lb and oz)

3. 4 lb 3 oz   (oz)   **8.** 36 oz   (lb and oz)

4. 3 tons 4 cwt (cwt)   **9.** 30 cwt   (tons and cwt)

5. 1 cwt 50 lb   (lb)   **10.** 120 lb   (cwt and lb)

ROUGH EQUIVALENCE BETWEEN METRIC AND IMPERIAL UNITS

If you shop in a supermarket you will find that nearly all prepacked goods (tinned foods, sugar, biscuits, etc.) are sold in grams or kilograms and nearly all fresh produce (meat, cheese, fruit, etc.) is sold in pounds and ounces. It is often useful to be able to convert, roughly, pounds into kilograms or grams into pounds. For a rough conversion it is good enough to say that

$$1 \text{ kg is about } 2 \text{ lb}$$

although one kilogram is slightly more than two pounds.

One metre is slightly longer than one yard but for a rough conversion it is good enough to say that

$$1 \text{ m is about } 1 \text{ yd}$$

Remember that the symbol $\approx$ means "is approximately equal to" so

$$1\,\text{kg} \approx 2\,\text{lb}$$

$$1\,\text{m} \approx 1\,\text{yd or }3\,\text{ft}$$

EXERCISE 8c   In questions 1 to 10, write the first unit roughly in terms of the unit in brackets:

> 5 kg   (lb)
>
> $$5\,\text{kg} \approx 5 \times 2\,\text{lb}$$
>
> $$\therefore\ 5\,\text{kg} \approx 10\,\text{lb}$$

> 10 ft   (m)
>
> $$10\,\text{ft} \approx 10 \div 3\,\text{m}$$
>
> $$\therefore\ 10\,\text{ft} \approx 3.3\,\text{m}\quad (\text{to 1 d.p.})$$

1. 3 kg   (lb)   **6.** 5 m   (ft)

2. 2 m   (ft)   **7.** 3.5 kg   (lb)

3. 4 lb   (kg)   **8.** 8 ft   (m)

4. 9 ft   (m)   **9.** 250 g   (oz)

5. 1.5 kg (lb)   **10.** 500 g   (lb)

In questions 11 to 16 use the approximation $5\,\text{miles} \approx 8\,\text{km}$ to convert the given number of miles into an approximate number of kilometres:

11. 10 miles   **13.** 15 miles   **15.** 75 miles

12. 20 miles   **14.** 100 miles   **16.** 40 miles

17. I buy a 5 lb bag of potatoes and two 1.5 kg bags of flour. What weight, roughly, in pounds do I have to carry?

18. A window is 6 ft high. Roughly, what is its height in metres?

19. I have a picture which measures 2 ft by 1 ft. Wood for framing it is sold by the metre. Roughly, what length of framing, in metres, should I buy?

20. In the supermarket I buy a 4 kg packet of sugar and a 5 lb bag of potatoes. Which is heavier?

21. In one catalogue a table cloth is described as measuring 4 ft by 8 ft. In another catalogue a different table cloth is described as measuring 1 m by 2 m. Which one is bigger?

22. The distance between London and Dover is about 70 miles. The distance between Calais and Paris is about 270 kilometres. Which is the greater distance?

23. A recipe requires 250 grams of flour. Roughly, how many ounces is this?

Converting from inches to centimetres and from centimetres to inches is often useful. For most purposes it is good enough to say that 1 inch $\approx 2\frac{1}{2}$ cm.

24. An instruction in an old knitting pattern says knit 6 inches. Mary has a tape measure marked only in centimetres. How many centimetres should she knit?

25. The instructions for repotting a plant say that it should go into a 10 cm pot. The flower pots that Tom has in his shed are marked 3 in, 4 in and 5 in. Which one should he use?

26. Peter Stuart wishes to extend his central heating which was installed several years ago in 1 in and $\frac{1}{2}$ in diameter copper tubing. The only new piping he can buy has diameters of 10 mm, 15 mm, 20 mm or 25 mm. Use the approximation 1 in ≈ 2.5 cm to determine which piping he should buy that would be nearest to

a) the 1 in pipes   b) the $\frac{1}{2}$ in pipes.

27. A carpenter wishes to replace a 6 in floorboard. The only sizes available are metric and have widths of 12 cm, 15 cm, 18 cm and 20 cm. Use the approximation 1 in ≈ 2.5 cm to determine which one he should buy.

28. A shop sells material at £10.50 per metre while the same material is sold in the local market at £9 per yard. Using 4 in ≈ 10 cm find which is cheaper.

9 INTRODUCING GEOMETRY

FRACTIONS OF A REVOLUTION

When the seconds hand of a clock starts at 12 and moves round until it stops at 12 again it has gone through one complete turn.

> One complete turn is called a revolution.

When the seconds hand starts at 12 and stops at 3 it has turned through $\frac{1}{4}$ of a revolution.

EXERCISE 9a What fraction of a revolution does the seconds hand of a clock turn through when:

it starts at 3 and stops at 12

$\frac{3}{4}$ of a revolution

it starts at 4 and stops at 8

$\frac{1}{3}$ of a revolution

1. it starts at 12 and stops at 9

2. it starts at 12 and stops at 6

123

3. it starts at 6 and stops at 9

4. it starts at 3 and stops at 9

5. it starts at 9 and stops at 12

6. it starts at 1 and stops at 7

7. it starts at 5 and stops at 11

8. it starts at 10 and stops at 4

9. it starts at 8 and stops at 8

10. it starts at 8 and stops at 11

11. it starts at 10 and stops at 2

12. it starts at 12 and stops at 4

13. it starts at 8 and stops at 5

14. it starts at 5 and stops at 2

15. it starts at 9 and stops at 5?

Where does the seconds hand stop if:

it starts at 12 and turns through $\frac{1}{4}$ of a revolution

It stops at 3.

16. it starts at 12 and turns through $\frac{1}{2}$ a turn

17. it starts at 12 and turns through $\frac{3}{4}$ of a turn

18. it starts at 6 and turns through $\frac{1}{4}$ of a turn

19. it starts at 9 and turns through $\frac{1}{2}$ a turn

20. it starts at 6 and turns through a complete turn

21. it starts at 9 and turns through $\frac{3}{4}$ of a turn

22. it starts at 12 and turns through $\frac{1}{3}$ of a turn

23. it starts at 12 and turns through $\frac{2}{3}$ of a turn

24. it starts at 9 and turns through a complete turn

25. it starts at 6 and turns through $\frac{1}{2}$ a turn?

BEARINGS

The four main compass directions are north, south, east and west.

If you stand facing north and turn clockwise through $\frac{1}{2}$ a revolution you are then facing south.

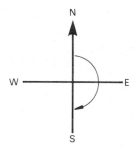

EXERCISE 9b

1. If you stand facing west and turn anticlockwise through $\frac{3}{4}$ of a revolution, in which direction are you facing?

2. If you stand facing south and turn clockwise through $\frac{1}{4}$ of a revolution, in which direction are you facing?

3. If you stand facing north and turn, in either direction, through a complete revolution, in which direction are you facing?

4. If you stand facing west and turn through $\frac{1}{2}$ a revolution, in which direction are you facing? Does it matter if you turn clockwise or anticlockwise?

5. If you stand facing south and turn through $1\frac{1}{2}$ revolutions, in which direction are you facing?

6. If you stand facing west and turn clockwise to face south what part of a revolution have you turned through?

7. If you stand facing north and turn clockwise to face west how much of a revolution have you turned through?

8. If you stand facing east and turn to face west what part of a revolution have you turned through?

ANGLES

When the hand of a clock moves from one position to another it has turned through an angle.

RIGHT ANGLES

A quarter of a revolution is called a *right angle*.

Half a revolution is two right angles.

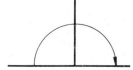

EXERCISE 9c   How many right angles does the seconds hand of a clock turn through when:

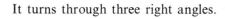

it starts at 3 and stops at 12

It turns through three right angles.

1. it starts at 6 and stops at 9

2. it starts at 3 and stops at 9

3. it starts at 12 and stops at 9

4. it starts at 3 and stops at 6

5. it starts at 12 and stops at 12

6. it starts at 8 and stops at 2

7. it starts at 9 and stops at 6

8. it starts at 7 and stops at 7?

How many right angles do you turn through if you:

9. face north and turn clockwise to face south

10. face west and turn clockwise to face north

11. face south and turn clockwise to face west

12. face north and turn anticlockwise to face east

13. face north and turn to face north again?

ACUTE, OBTUSE AND REFLEX ANGLES

Any angle that is smaller than a right angle is called an *acute angle*.

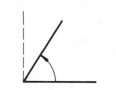

Any angle that is greater than one right angle and less than two right angles is called an *obtuse angle*.

Any angle that is greater than two right angles is called a *reflex angle*.

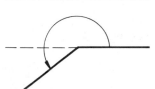

EXERCISE 9d   What type of angle is each of the following?

1.

2.

3.

4.

5.

6.

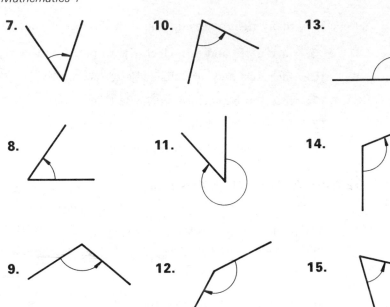

DEGREES

One complete revolution is divided into 360 parts. Each part is called a *degree*. 360 degrees is written 360°.

360 seems a strange number of parts to have in a revolution but it is a good number because so many whole numbers divide into it exactly. This means that there are many fractions of a revolution that can be expressed as an exact number of degrees.

EXERCISE 9e **1.** How many degrees are there in half a revolution?

2. How many degrees are there in one right angle?

3. How many degrees are there in three right angles?

How many degrees has the seconds hand of a clock turned through when it moves from 6 to 9?

It has turned through 90°.

How many degrees has the seconds hand of a clock turned through when it moves from:

4.	12 to 6	**8.**	9 to 6	**12.**	8 to 5
5.	3 to 6	**9.**	2 to 5	**13.**	4 to 10
6.	6 to 3	**10.**	7 to 11	**14.**	5 to 8
7.	9 to 3	**11.**	1 to 10	**15.**	6 to 12?

How many degrees has the seconds hand of a clock turned through when it moves from:

6 to 8

Either $\frac{2}{3}$ of $90° = 60°$

or $\frac{2}{12}$ of $360° = 60°$

16.	8 to 9	**21.**	4 to 5
17.	10 to halfway between 11 and 12	**22.**	7 to 11
18.	6 to 10	**23.**	5 to 6
19.	1 to 3	**24.**	7 to 9
20.	3 to halfway between 4 and 5	**25.**	11 to 3

26.	3 to 10	**31.**	8 to 3
27.	2 to 8	**32.**	7 to 5
28.	10 to 8	**33.**	10 to 5
29.	12 to 11	**34.**	11 to 4
30.	9 to 2	**35.**	2 to 9?

USING A PROTRACTOR TO MEASURE ANGLES

A protractor looks like this:

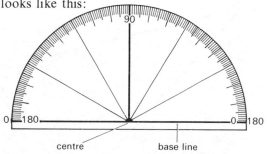

It has a straight line at or near the straight edge. This line is called the *base line*.

The *centre* of the base line is marked.

The protractor has two scales, an inside one and an outside one.

To measure the size of this angle, first decide whether it is acute or obtuse.

This is an acute angle because it is *less* than 90°.

Next place the protractor on the angle as shown.

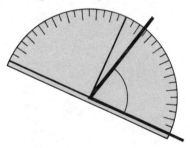

One arm of the angle is on the base line.

The vertex (point) of the angle is at the centre of the base line.

Choose the scale that starts at 0° on the arm on the base line. Read off the number where the other arm cuts this scale.

Check with your estimate to make sure that you have chosen the right scale.

EXERCISE 9f Measure the following angles (if necessary, turn the page to a convenient position):

1.   **2.**

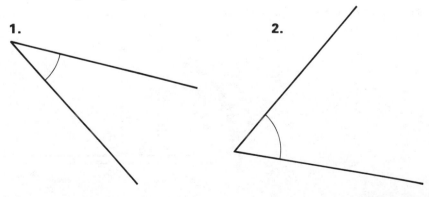

3.

7.

8.

4.

5.

9.

6.

10.

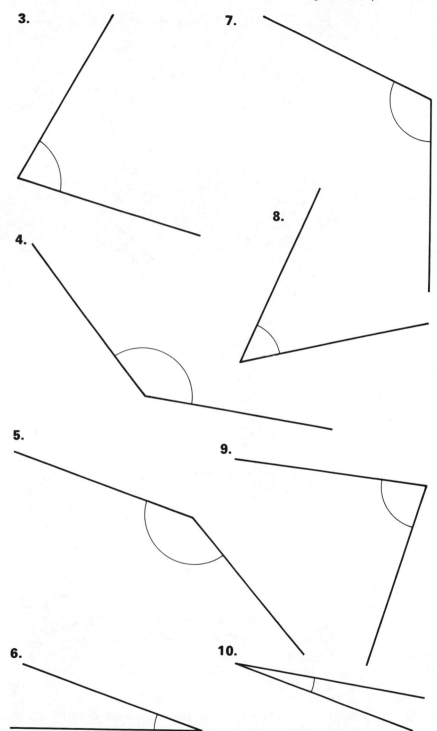

In questions 11 to 15 write down the size of the angle marked with a letter:

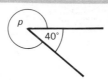

Angle *p* and 40° make 360°.

So angle *p* is 360° − 40° = 320°.

11. 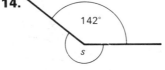 **12.** **13.**

14. 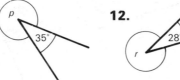 **15.**

Find the following angle:

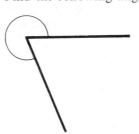

This is a reflex angle and it is bigger than 3 right angles, i.e. it is greater than 270°.

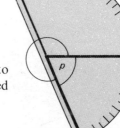

To find this angle, we need to measure the smaller angle, marked *p*.

Angle *p* is 68° so the reflex angle is 360° − 68° = 292°.

Find the following angles:

16.

17.

18.

19.

20.

21.

22. Draw a reflex angle. Now find its size. Change books with your neighbour and each check the other's measurement.

MIXED QUESTIONS

EXERCISE 9g   Use a clock diagram to draw the angle that the *minute* hand of a clock turns through in the following times. In each question write down the size of the angle in degrees:

1. 5 minutes   **3.** 15 minutes   **5.** 25 minutes

2. 10 minutes   **4.** 20 minutes   **6.** 30 minutes

The seconds hand of a clock starts at 12. Which number is it pointing to when it has turned through an angle of:

7. 90°   **11.** 150°   **15.** 420°   **19.** 540°

8. 60°   **12.** 270°   **16.** 180°   **20.** 240°

9. 120°   **13.** 30°   **17.** 450°   **21.** 390°

10. 360°   **14.** 300°   **18.** 210°   **22.** 720°?

If you start by facing north and turn clockwise, draw a sketch to show roughly the direction in which you are facing if you turn through:

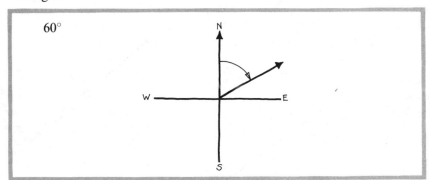

60°

23. 45°   **26.** 50°   **29.** 20°   **32.** 10°

24. 70°   **27.** 200°   **30.** 100°   **33.** 80°

25. 120°   **28.** 300°   **31.** 270°   **34.** 250°

Estimate the size, in degrees, of each of the following angles:

35.   **36.**    **37.**

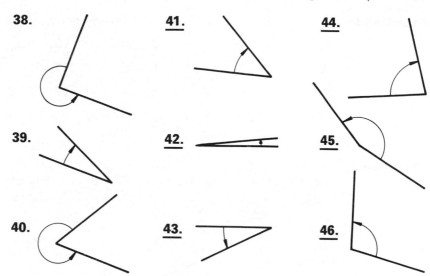

38. 39. 40. 41. 42. 43. 44. 45. 46.

Draw the following angles as well as you can by estimating, i.e. without using a protractor. Use a clockface if it helps. Then measure your angles with a protractor.

47.	45°	**50.**	30°	**53.**	150°	**56.**	20°	**59.**	330°
48.	90°	**51.**	60°	**54.**	200°	**57.**	5°	**60.**	95°
49.	120°	**52.**	10°	**55.**	290°	**58.**	170°	**61.**	250°

DRAWING ANGLES USING A PROTRACTOR

To draw an angle of 120° start by drawing one arm and mark the vertex.

Place your protractor as shown in the diagram. Make sure that the vertex is at the centre of the base line.

Choose the scale that starts at 0° on your drawn line and mark the paper next to the 120° mark on the scale.

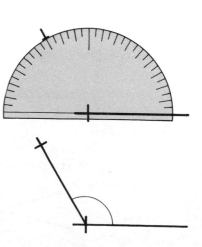

Remove the protractor and join your mark to the vertex.

Now look at your angle: does it look the right size?

EXERCISE 9h   Use your protractor to draw the following angles accurately:

1. 25°	**4.** 160°	**7.** 110°	**10.** 125°	**13.** 105°
2. 37°	**5.** 83°	**8.** 49°	**11.** 175°	**14.** 136°
3. 55°	**6.** 15°	**9.** 65°	**12.** 72°	**15.** 85°

Change books with your neighbour and measure each other's angles as a check on accuracy.

EXERCISE 9i   In questions 1 and 2 first measure the angle marked *r*. Then estimate the size of the angle marked *s*. Check your estimate by measuring angle *s*.

1.

2.

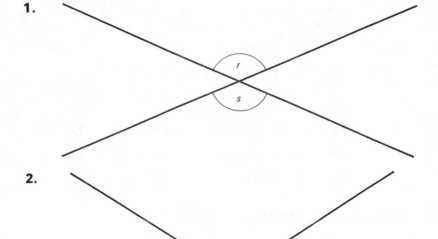

3.   Draw some more similar diagrams and repeat questions 1 and 2.

In each of the following questions, write down the size of the angle marked *t*, without measuring it:

4.

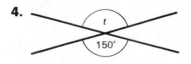

5.

6.

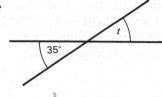

8.

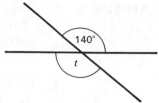

7.

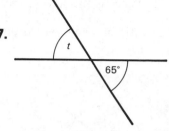

9.

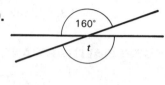

VERTICALLY OPPOSITE ANGLES

When two straight lines cross, four angles are formed.

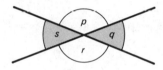

The two angles that are opposite each other are called *vertically opposite angles*. After working through the last exercise you should now be convinced that

> vertically opposite angles are equal

i.e. $p = r$ and $s = q$.

ANGLES ON A STRAIGHT LINE

The seconds hand of a clock starts at 9 and stops at 11 and then starts again and finally stops at 3.

Altogether the seconds hand has turned through half a revolution, so $p+q = 180°$.

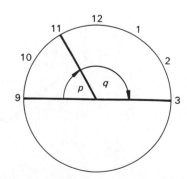

EXERCISE 9j   **1.**   Draw a diagram showing the two angles that you turn through if you start by facing north and then turn clockwise through 60°, stop for a moment and then continue turning until you are facing south. What is the sum of these two angles?

2.   Draw a clock diagram to show the two angles turned through by the seconds hand if it is started at 2, stopped at 6, started again and finally stopped at 8. What is the sum of these two angles?

3.   Draw an angle of 180°, without using your protractor.

SUPPLEMENTARY ANGLES

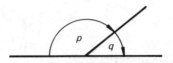

Angles on a straight line add up to 180°.

Two angles that add up to 180° are called supplementary angles.

EXERCISE 9k   In questions 1 to 12 calculate the size of the angle marked with a letter:

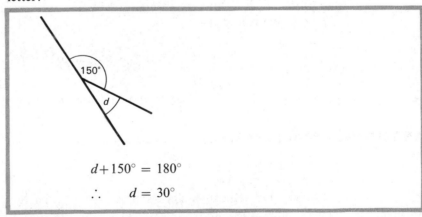

$$d + 150° = 180°$$
$$\therefore \quad d = 30°$$

1.

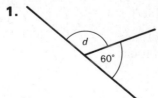

2.

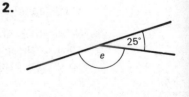

3.

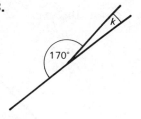

170°   k

4.

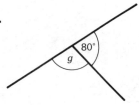

80° g

5.

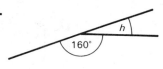

h   160°

6.

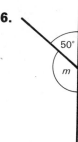

50° m

7.

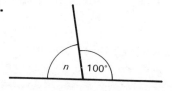

n   100°

<u>**8.**</u>

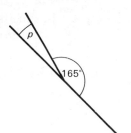

p   165°

<u>**9.**</u>

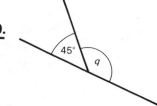

45° q

<u>**10.**</u>

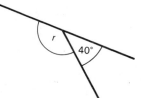

r   40°

<u>**11.**</u>

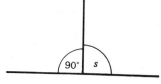

90° s

<u>**12.**</u>

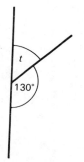

t   130°

In questions 13 to 18 write down the pairs of angles that are supplementary:

13.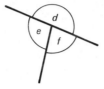

14.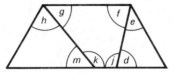

15.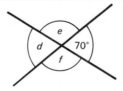

16.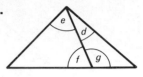

17.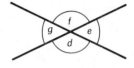

18.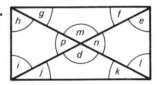

In questions 19 to 26 calculate the size of the angles marked with a letter:

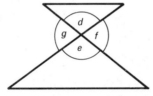

d and 70° are equal (they are vertically opposite)

∴ *d* = 70°

e and 70° add up 180° (they are angles on a straight line)

∴ *e* = 110°

f and *e* are equal (they are vertically opposite)

∴ *f* = 110°

19.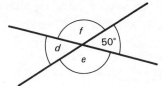

23.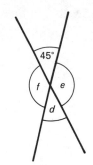

20.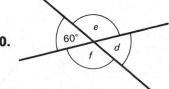

24.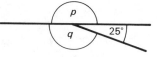

21.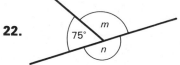

25.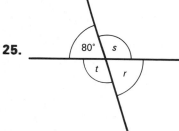

22.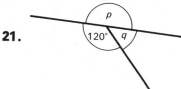

26.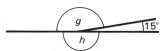

ANGLES AT A POINT

When several angles make a complete revolution they are called *angles at a point*.

Angles at a point add up to 360°.

EXERCISE 9I   In questions 1 to 10 find the size of the angle marked with a letter:

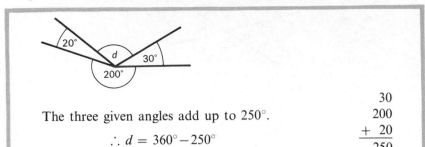

The three given angles add up to 250°.

$$\therefore d = 360° - 250°$$

$$d = 110°$$

$$\begin{array}{r} 30 \\ 200 \\ + \ 20 \\ \hline 250 \end{array}$$

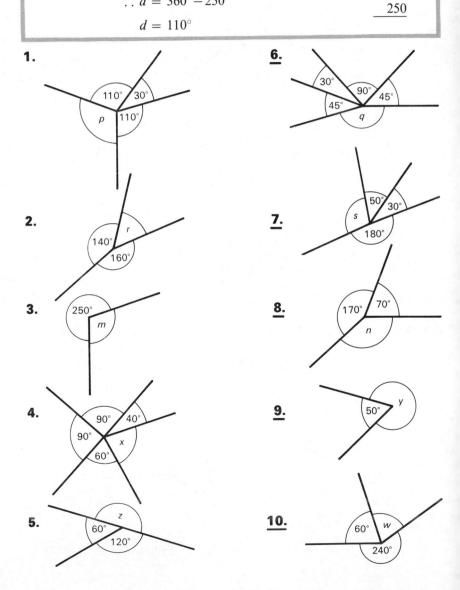

1.

2.

3.

4.

5.

6.

7.

8.

9.

10.

PROBLEMS

EXERCISE 9m **1.** Find each of the equal angles marked *s*.

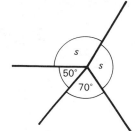

2. The angle marked *f* is twice the angle marked *g*. Find angles *f* and *g*.

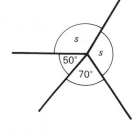

3. Find each of the equal angles marked *d*.

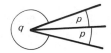

4. Each of the equal angles marked *p* is 25°. Find the reflex angle *q*.

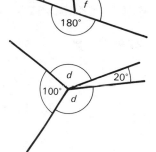

5. Each of the equal angles marked *d* is 30°. Angle *d* and angle *e* are supplementary. Find angles *e* and *f*. (An angle marked with a square is a right angle.)

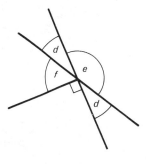

6. Angle *s* is twice angle *t*. Find angle *r*.

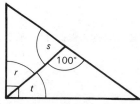

7. The angle marked *d* is 70°. Find angle *e*.

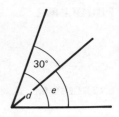

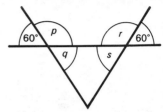

8. Find the angles marked *p*, *q*, *r* and *s*.

MIXED EXERCISES

EXERCISE 9n **1.** What angle does the minute hand of a clock turn through when it moves from 1 to 9?

2. Draw an angle of 50°.

3. Estimate the size of this angle:

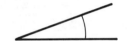

4. Write down the size of the angle marked *p*.

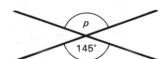

5. Write down the size of the angle marked *s*.

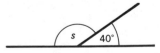

6. Find each of the equal angles marked *e*.

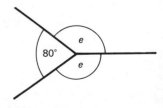

EXERCISE 9p **1.** What angle does the minute hand of a clock turn through when it moves from 10 to 6?

2. If you start facing north and turn clockwise through an angle of 270°, in which direction are you then facing?

3.   Measure the angle marked *q*.

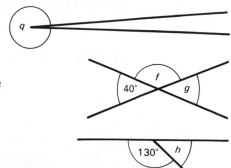

4.   Write down the sizes of the
angles marked *f* and *g*.

5.   Write down the size of the
angle marked *h*.

6.   Angles *p* and *q* are supplementary. Angle *p* is five times the size
of angle *q*. What is the size of angle *q*?

10 SYMMETRY

LINE SYMMETRY

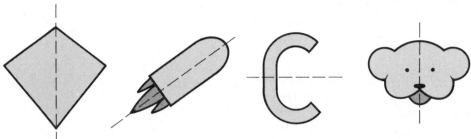

The four shapes above are *symmetrical*. If they were folded along the broken line, one half of the drawing would fit exactly over the other half.

Fold a piece of paper in half and cut a shape from the folded edge. When unfolded, the resulting shape is symmetrical. The fold line is the *axis of symmetry*.

EXERCISE 10a Some of the shapes below have one axis of symmetry and some have none. State which of the drawings 1 to 6 have an axis of symmetry.

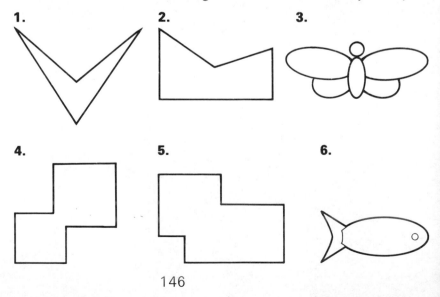

1.

2.

3.

4.

5.

6.

Copy the following drawings on squared paper and complete them so that the broken line is the axis of symmetry:

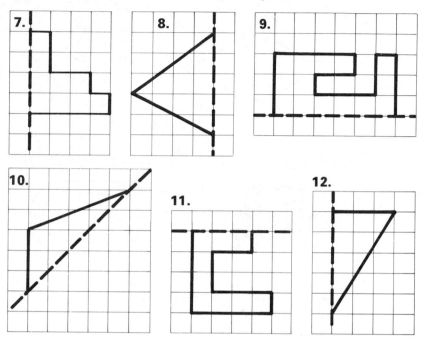

TWO AXES OF SYMMETRY

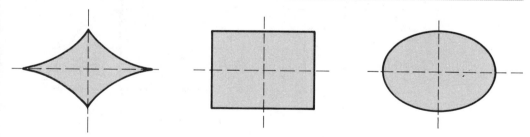

In these shapes there are two lines along which it is possible to fold the paper so that one half fits exactly over the other half.

Fold a piece of paper twice, cut a shape as shown and unfold it. The resulting shape has two axes of symmetry.

EXERCISE 10b How many axes of symmetry are there in each of the following shapes?

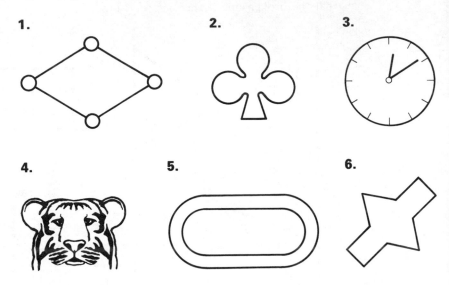

1.

2.

3.

4.

5.

6.

Copy the following drawings on squared paper and complete them so that the two broken lines are the two axes of symmetry.

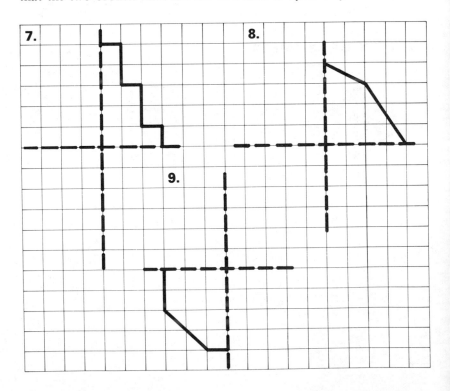

7.

8.

9.

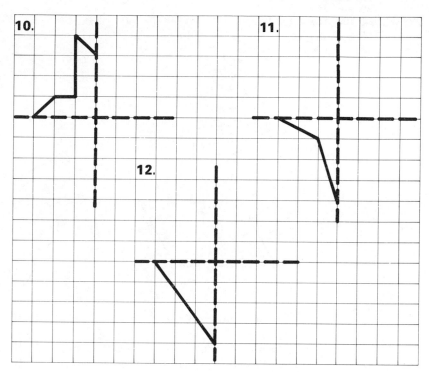

THREE OR MORE AXES OF SYMMETRY

It is possible to have more than two lines of symmetry.

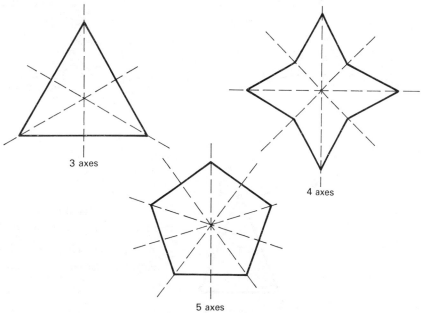

3 axes

4 axes

5 axes

EXERCISE 10c How many axes of symmetry are there for each of the following shapes?

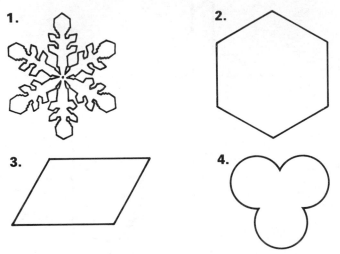

1.

2.

3.

4.

5. Fold a square piece of paper twice then fold it a third time along the broken line. Cut a shape, simple or complicated, and unfold the paper. How many axes of symmetry does it have?

6.

Copy the triangle on squared paper and mark in the axis of symmetry.

A triangle with an axis of symmetry is called an *isosceles triangle*.

7.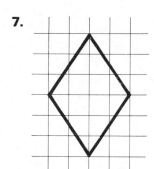

Copy the quadrilateral on squared paper and mark in the two axes of symmetry.

This quadrilateral (which has four equal sides) is called a *rhombus*.

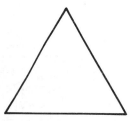

8. Trace the triangle. Draw in its axes of symmetry. Measure its three sides.

This triangle is called an *equilateral triangle*.

ROTATIONAL SYMMETRY (S-SYMMETRY)

These shapes have a different type of symmetry. They cannot be folded in half but can be turned or rotated about a centre point (marked with ×) and still look the same.

EXERCISE 10d 1. Lay a piece of tracing paper over any one of the shapes above, trace it and turn it about the cross until it fits over the shape again.

Which of the following shapes have rotational symmetry?

2. 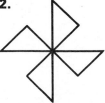   **3.** 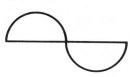   **4.**

5. 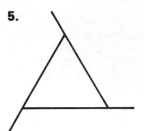   **6.** 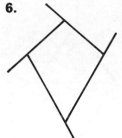   **7.**

Some shapes have both line symmetry and rotational symmetry:

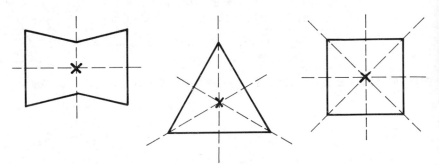

8. Sketch the capital letters of the alphabet. Mark any axes of symmetry and the centre of rotation if it exists.

For instance, draw H.

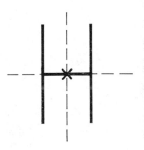

9. Which of the shapes in Exercise 10c have rotational symmetry?

11 TRIANGLES AND ANGLES

CONSTRUCTIONS

When a new object, for example a new car, is designed there are many jobs that have to be done before it can be made. One of these jobs is to make accurate drawings of the parts. This is called technical drawing.

To draw accurately you need

- a *sharp* pencil
- a ruler
- a pair of compasses
- a protractor

USING A PAIR OF COMPASSES

Using a pair of compasses is not easy: it needs practice. Draw several circles. Make some of them small and some large. You should not be able to see the place at which you start and finish.

Now try drawing the daisy pattern below.

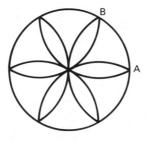

Draw a circle of radius 5 cm. Keeping the radius the same, put the point of the compasses at A and draw an arc to meet the circle in two places, one of which is B. Move the point to B and repeat. Carry on moving the point of your compasses round the circle until the pattern is complete.

Repeat the daisy pattern but this time draw complete circles instead of arcs.

There are some more patterns using compasses in Exercise 11k on page 171.

DRAWING STRAIGHT LINES OF A GIVEN LENGTH

To draw a straight line that is 5 cm long, start by using your ruler to draw a line that is *longer* than 5 cm.

Then mark a point on this line near one end as shown.
Label it A.

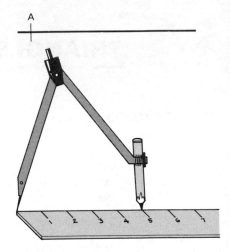

Next use your compasses to measure 5 cm on your ruler.

Then put the point of the compasses on the line at A and draw an arc to cut the line as shown.

The length of line between A and B should be 5 cm. Measure it with your ruler.

EXERCISE 11a Draw, as accurately as you can, straight lines of the following lengths:

1.	6 cm	**3.**	12 cm	**5.**	8.5 cm	**7.**	4.5 cm
2.	2 cm	**4.**	9 cm	**6.**	3.5 cm	**8.**	6.8 cm

TRIANGLES

A triangle has three sides and three angles.

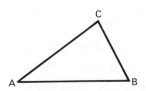

The corners of the triangle are called vertices. (One corner is called a vertex.) So that we can refer to one particular side, or to one particular angle, we label the vertices using capital letters. In the diagram above we used the letters A, B and C so we can now talk about "the triangle ABC" or "△ABC".

The side between A and B is called "the side AB" or AB.

The side between A and C is called "the side AC" or AC.

The side between B and C is called "the side BC" or BC.

The angle at the corner A is called "angle A" or $\widehat{A}$ for short.

EXERCISE 11b **1.** Write down the name of the side which is 4 cm long.

Write down the name of the side which is 2 cm long.

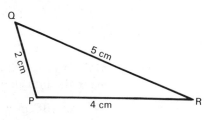

2. Write down the name of

a) the side which is 2.5 cm long
b) the side which is 2 cm long
c) the angle which is 70°

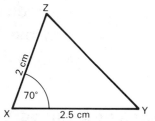

In the following questions, draw a rough copy of the triangle and mark the given measurements on your drawing:

3. In △ABC, AB = 4 cm, $\widehat{B}$ = 60°, $\widehat{C}$ = 50°.

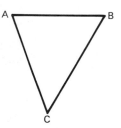

4. In △DEF, $\widehat{E}$ = 90°, $\widehat{F}$ = 70°, EF = 3 cm.

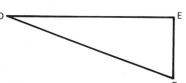

5. In △LMN, $\widehat{L}$ = 100°, $\widehat{N}$ = 30°, NL = 2.5 cm.

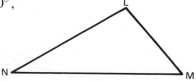

6. In △FGH, FG = 3.5 cm,
GH = 3 cm, $\hat{H}$ = 35°.

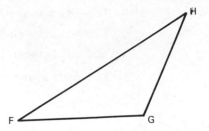

Make a rough drawing of the following triangles. Label each one and mark the measurements given:

7. △ABC in which AB = 10 cm, BC = 8 cm and $\hat{B}$ = 60°.

8. △PQR in which $\hat{P}$ = 90°, $\hat{Q}$ = 30° and PQ = 6 cm.

9. △DEF in which DE = 8 cm, $\hat{D}$ = 50° and DF = 6 cm.

10. △XYZ in which XY = 10 cm, $\hat{X}$ = 30° and $\hat{Y}$ = 80°.

ANGLES OF A TRIANGLE

Draw a large triangle of any shape. Use a straight edge to draw the sides. Measure each angle in this triangle, turning your page to a convenient position when necessary. Add up the sizes of the three angles.

Draw another triangle of a different shape. Again measure each angle and then add up their sizes.

Now try this: on a piece of paper draw a triangle of any shape and cut it out. Next tear off each corner and place the three corners together.

They should look like this:

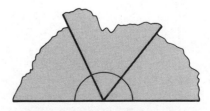

The three angles of a triangle add up to 180°.

EXERCISE 11c Find the size of angle A (an angle marked with a square is a right angle):

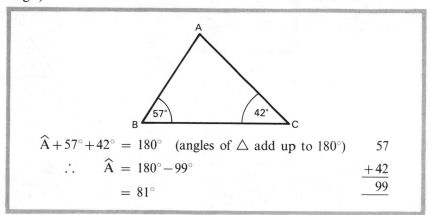

$\widehat{A} + 57° + 42° = 180°$ (angles of △ add up to 180°)

$\therefore \quad \widehat{A} = 180° - 99°$

$\quad = 81°$

$\begin{array}{r} 57 \\ +42 \\ \hline 99 \\ \hline \end{array}$

1.

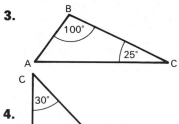

6.

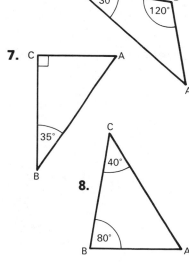

7.

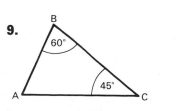

2.

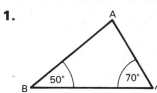

3.

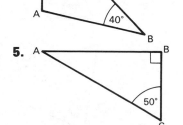

8.

4.

9.

5.

10.

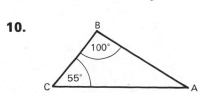

11.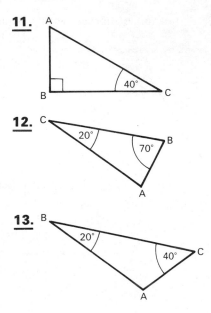

12.

13.

14.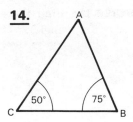

15.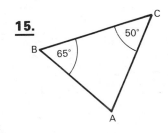

PROBLEMS

Reminder: Vertically opposite angles are equal.

Angles on a straight line add up to 180°.

You will need these facts in the next exercise.

EXERCISE 11d In each question make a rough copy of the diagram and mark the sizes of the angles that you are asked to find:

1. Find angles *d* and *f*.

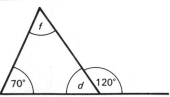

2. Find angles *s* and *t*.

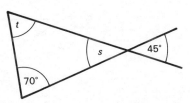

3. Find each of the equal angles *x*.

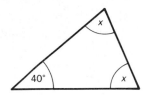

4. Find angles *p* and *q*.

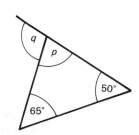

5. Find angles *s* and *t*.

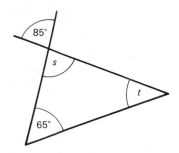

6. Find each of the equal angles *g*.

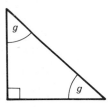

7. Find each of the equal angles *x*.

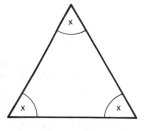

8. Angle *h* is twice angle *j*.
Find angles *h* and *j*.

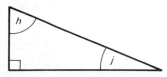

9. Find each of the equal angles q and angle p.

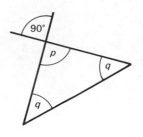

CONSTRUCTING TRIANGLES GIVEN ONE SIDE AND TWO ANGLES

If we are given enough information about a triangle we can make an accurate drawing of that triangle. The mathematical word for "make an accurate drawing of" is "construct".

For example: construct $\triangle ABC$ in which $AB = 7\,cm$, $\widehat{A} = 30°$ and $\widehat{B} = 40°$.

First make a rough sketch of $\triangle ABC$ and put all the given measurements in your sketch.

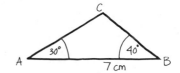

Next draw the line AB making it 7 cm long. Label the ends.

Then use your protractor to make an angle of 30° at A.

Next make an angle of 40° at B. If necessary extend the arms of the angles until they cross; this is the point C.

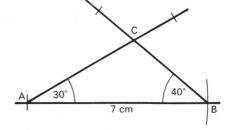

We can calculate $\widehat{C}$ because $\widehat{A}+\widehat{B}+\widehat{C} = 180°$ so $\widehat{C} = 110°$. Now as a check we can measure $\widehat{C}$ in our construction.

EXERCISE 11e Construct the following triangles; calculate the third angle in each triangle and then measure this angle to check the accuracy of your construction.

    1. $\triangle ABC$ in which $AB = 8\,cm$, $\widehat{A} = 50°$, $\widehat{B} = 20°$

    2. $\triangle PQR$ in which $QR = 5\,cm$, $\widehat{Q} = 30°$, $\widehat{R} = 90°$

3. △DEF in which EF = 6 cm, $\hat{E}$ = 50°, $\hat{F}$ = 60°

4. △XYZ in which YZ = 5.5 cm, $\hat{Y}$ = 100°, $\hat{Z}$ = 40°

5. △UVW in which $\hat{V}$ = 35°, VW = 5.5 cm, $\hat{W}$ = 75°

6. △FGH in which $\hat{F}$ = 55°, $\hat{G}$ = 70°, FG = 4.5 cm

7. △KLM in which KM = 10 cm, $\hat{K}$ = 45°, $\hat{M}$ = 45°

8. △BCD in which $\hat{B}$ = 100°, BC = 8.5 cm, $\hat{C}$ = 45°

9. △GHI in which GI = 7 cm, $\hat{G}$ = 25°, $\hat{I}$ = 45°

10. △JKL in which $\hat{J}$ = 50°, JL = 6.5 cm, $\hat{L}$ = 35°

CONSTRUCTING TRIANGLES GIVEN TWO SIDES AND THE ANGLE BETWEEN THE TWO SIDES

To construct △PQR in which PQ = 4.5 cm, PR = 5.5 cm and $\hat{P}$ = 35°, first draw a rough sketch of △PQR and put in all the measurements that you are given.

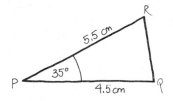

Draw one of the sides whose length you know; we will draw PQ.

Now using your protractor make an angle of 35° at P. Make the arm of the angle quite long.

Next use your compasses to measure the length of PR on your ruler.

Then with the point of your compasses at P, draw an arc to cut the arm of the angle. This is the point R.

Now join R and Q.

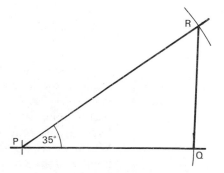

162   *ST(P) Mathematics 1*

EXERCISE 11f Construct each of the following triangles and measure the third side:

1. △ABC in which AB = 5.5 cm, BC = 6.5 cm, $\hat{B} = 40°$

2. △PQR in which PQ = 6 cm, QR = 8 cm, $\hat{Q} = 35°$

3. △XYZ in which XZ = 4.5 cm, YZ = 6.5 cm, $\hat{Z} = 70°$

4. △DEF in which DE = 6 cm, $\hat{E} = 50°$, EF = 11 cm

5. △HJK in which HK = 4.2 cm, $\hat{H} = 45°$, HJ = 5.3 cm

6. △ABC in which AC = 6.3 cm, $\hat{C} = 48°$, CB = 5.1 cm

7. △XYZ in which $\hat{Y} = 65°$, XY = 3.8 cm, YZ = 4.2 cm

8. △PQR in which $\hat{R} = 52°$, RQ = 5.8 cm, PR = 7 cm

9. △LMN in which $\hat{N} = 73°$, LN = 4.1 cm, MN = 6.3 cm

10. △ABC in which AC = 5.2 cm, BA = 7.3 cm, $\hat{A} = 56°$

CONSTRUCTING TRIANGLES GIVEN THE LENGTHS OF THE THREE SIDES

To construct △XYZ in which XY = 5.5 cm, XZ = 3.5 cm, YZ = 6.5 cm first draw a rough sketch of the triangle and put in all the given measurements.

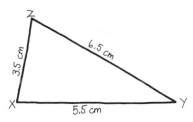

Next draw one side; we will draw XY.

Then with your compasses measure the length of XZ from your ruler. With the point of your compasses at X draw a wide arc.

Next use your compasses to measure the length of YZ from your ruler. Then with your compasses point at Y draw another large arc to cut the first arc. Where the two arcs cross is the point Z. Join ZX and ZY.

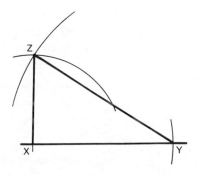

EXERCISE 11g Construct the following triangles:

1. △ABC in which AB = 7 cm, BC = 8 cm, AC = 12 cm

2. △PQR in which PQ = 4.5 cm, PR = 6 cm, QR = 8 cm

3. △XYZ in which XZ = 10.4 cm, XY = 6 cm, YZ = 9.6 cm

4. △DEF in which DE = 8 cm, DF = 10 cm, EF = 6 cm

5. △ABC in which AB = 7.3 cm, BC = 6.1 cm, AC = 4.7 cm

6. △DEF in which DE = 10.4 cm, EF = 7.4 cm, DF = 8.2 cm

7. △PQR in which PQ = 8.8 cm, QR = 6.6 cm, PR = 11 cm

8. △LMN in which LN = 7 cm, NM = 5.3 cm, LM = 6.1 cm

9. △XYZ in which XY = 12 cm, YZ = 5 cm, XZ = 13 cm

10. △ABC in which AB = 5.5 cm, BC = 6 cm, AC = 6.5 cm

EXERCISE 11h Construct the following triangles. Remember to draw a rough diagram of the triangle first and then decide which method you need to use:

1. △ABC in which AB = 7 cm, $\hat{A}$ = 30°, $\hat{B}$ = 50°

2. △PQR in which PQ = 5 cm, QR = 4 cm, RP = 7 cm

3. △BCD in which $\hat{B}$ = 60°, BC = 5 cm, BD = 4 cm

4. △WXY in which WX = 5 cm, XY = 6 cm, $\hat{X}$ = 90°

5. △KLM in which KL = 6.4 cm, LM = 8.2 cm, KM = 12.6 cm

6. △ABC in which $\hat{A}$ = 45°, AC = 8 cm, $\hat{C}$ = 110°

7. △DEF in which $\hat{E}$ = 125°, DE = 4.5 cm, FE = 5.5 cm

8. △PQR in which $\hat{P}$ = 72°, $\hat{R}$ = 53°, PR = 5.1 cm

9. △XYZ in which XY = 4 cm, YZ = 6 cm, XZ = 9 cm

10. △CDE in which CD = DE = 6 cm, $\hat{D}$ = 60°

11. Try to construct a triangle ABC in which $\hat{A}$ = 30°, AB = 5 cm, BC = 3 cm.

12. Construct two triangles which fit the following measurements: △PQR in which $\hat{P}$ = 60°, PQ = 6 cm, QR = 5.5 cm.

13. Construct △ABC in which $\hat{A}$ = 120°, AB = 4 cm, BC = 6 cm. Can you construct more than one triangle that fits these measurements?

QUADRILATERALS

A quadrilateral has four sides. These shapes are examples of quadrilaterals:

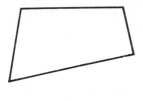 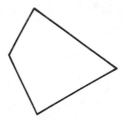 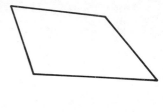

The following diagrams are also quadrilaterals, but each one is a "special" quadrilateral with its own name:

 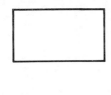   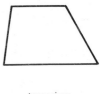

square   rectangle   parallelogram   rhombus   trapezium

Draw yourself a large quadrilateral, but do not make it one of the special cases. Measure each angle and then add up the sizes of the four angles.

Do this again with another three quadrilaterals.

Now try this: on a piece of paper draw a quadrilateral. Tear off each corner and place the vertices together. It should look like this:

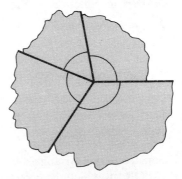

The sum of the four angles of a quadrilateral is 360°.

This is true of any quadrilateral whatever its shape or size.

EXERCISE 11i Make a rough copy of the following diagrams and mark on your diagram the sizes of the required angles. You can also write in the sizes of any other angles that you may need to find.

In questions 1 to 10 find the size of the angle marked d:

1.

130°
d

6.

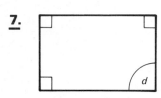

140°
d

2.

100°
70°
110°
d

7.

d

8.

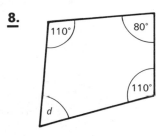

110°
80°
110°
d

3.

d
120°
70°
60°

9.

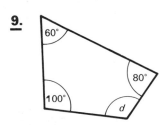

60°
80°
100°
d

4.

100°
120°
d

10.

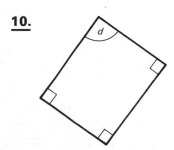

d

5.

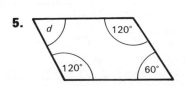

d
120°
120°
60°

11. Find each of the equal angles *d*.

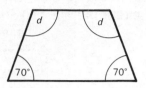

12. Find each of the equal angles *d*.

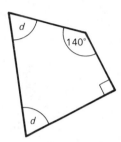

13. Angle *e* is twice angle *d*. Find angles *d* and *e*.

14. Find angles *d* and *e*.

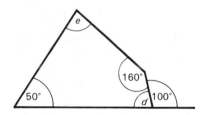

15. Find each of the equal angles *e*.

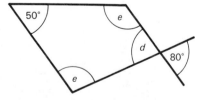

16. Angles *d* and *e* are supplementary. Find each of the equal angles *e*.

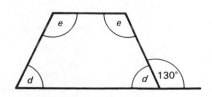

SOME SPECIAL TRIANGLES: EQUILATERAL AND ISOSCELES

A triangle in which all three sides are the same length is called an *equilateral triangle*.

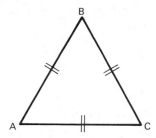

Construct an equilateral triangle in which the sides are each of length 6 cm. Label the vertices A, B and C.

On a separate piece of paper construct a triangle of the same size and cut it out. Label the angles A, B and C inside the triangle.

Place it on your first triangle. Now turn it round and it should still fit exactly. What do you think this means about the three angles? Measure each angle in the triangle.

In an equilateral triangle all three sides are the same length and each of the three angles is 60°.

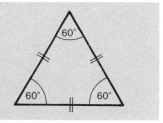

A triangle in which two sides are equal is called an *isosceles triangle*.

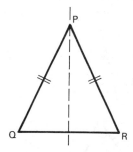

On a piece of paper construct an isosceles triangle PQR in which PQ = 8 cm, PR = 8 cm and $\hat{P} = 80°$. Cut it out and fold the triangle through P so that the corners at Q and R meet. You should find that $\hat{Q} = \hat{R}$. (The fold line is a line of symmetry.)

In an isosceles triangle two sides are equal and the two angles at the base of the equal sides are equal.

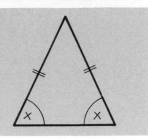

EXERCISE 11j   In questions 1 to 10 make a rough sketch of the triangle and mark angles that are equal:

1.
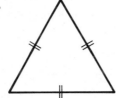

2.
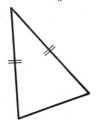

3.
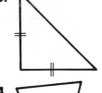

4.
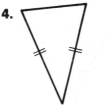

5.

6.

7.
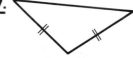

8.

9.

10.
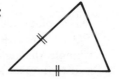

In questions 11 to 22 find angle *d*:

11.
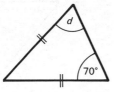

12.
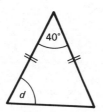

13.
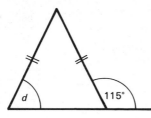

14.
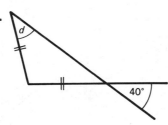

15.
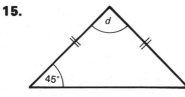

16.
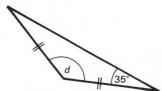

17.
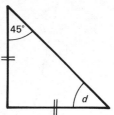

18.
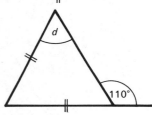

19.
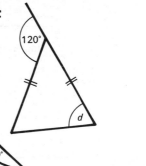

20.
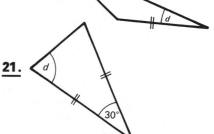

21.
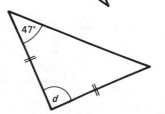

22.

In questions 23 to 26 mark the equal sides:

23.

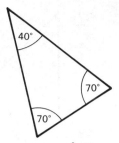

25.

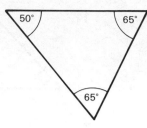

24.

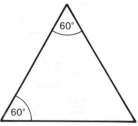

26.

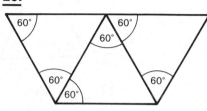

In questions 27 to 32 find angles *d* and *e*:

27.

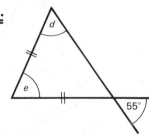

30.

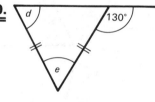

31.

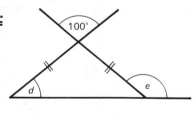

28.

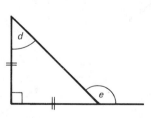

29.

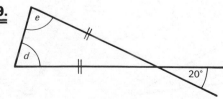

32.

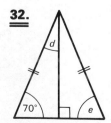

EXERCISE 11k The patterns below are made using a pair of compasses. Try copying them. Some instructions are given which should help.

1.

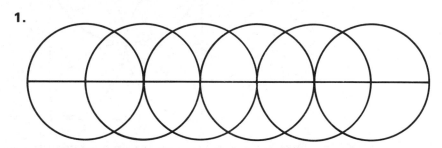

Draw a straight line. Open your compasses to a radius of 3 cm and draw a circle with its centre on the line. Move the point of the compasses 3 cm along the line and draw another circle. Repeat as often as you can.

2. Draw a square of side 4 cm. Open your compasses to a radius of 4 cm and with the point on one corner of the square draw an arc across the square. Repeat on the other three corners.

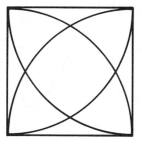

Try the same pattern, but leave out the sides of the square; just mark the corners. A block of four of these looks good.

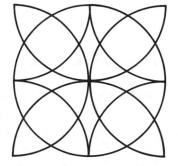

3. Draw a square of side 8 cm. Mark the midpoint of each side. Open your compasses to a radius of 4 cm, and with the point on the middle of one side of the square, draw an arc. Repeat at the other three midpoints.

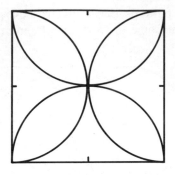

4. On a piece of paper construct an equilateral triangle of side 4 cm.

Construct an equilateral triangle, again of side 4 cm, on each of the three sides of the first triangle.

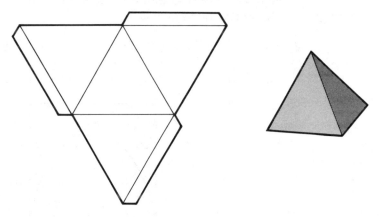

Cut out the complete diagram. Fold the outer triangles up so that the corners meet. Stick the edges together using the tabs. You have made a tetrahedron. (These make good Christmas tree decorations if made out of foil-covered paper.)

MIXED EXERCISES

EXERCISE 11I 1. Find the size of the angle marked x.

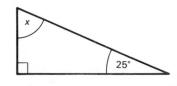

2.

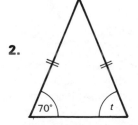

Find the size of the angle marked t.

3. Find the size of the angle marked *y*.

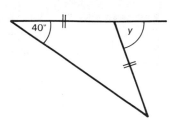

4. Construct △ABC in which AB = 6 cm, BC = 4 cm and B̂ = 40°. Measure AC.

5. Construct △ABC in which Â = 90°, AB = 6 cm, AC = 8 cm. Measure BC.

EXERCISE 11m 1. Find the size of the angles marked *p* and *q*.

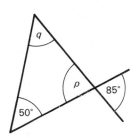

2. Find the size of the angles marked *x* and *y*.

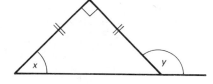

3. Find the size of the angles marked *u* and *v*.

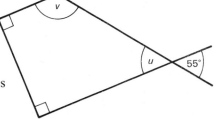

4. Construct △ABC in which Â = 50°, B̂ = 60° and BC = 5 cm. Be careful: this question needs some calculation before you can construct △ABC.

5. Construct the isosceles triangle ABC in which AB = BC = 6 cm and one of the base angles is 70°.

EXERCISE 11n

1. All three sides of the large triangle are equal. Find angles *r* and *s*.

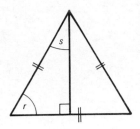

2. Find angles *x*, *y* and *z*.

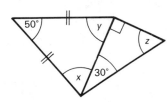

3. Find angles *f* and *g*.

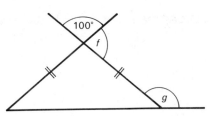

4. Construct an isosceles triangle in which the equal sides are of length 5 cm and one of the base angles is 45°.

5. Construct a quadrilateral ABCD in which AB = 5 cm, $\hat{A}$ = 70°, AD = 3 cm, BC = 5 cm, DC = 5 cm. Measure all four angles in the quadrilateral and find their sum.

12 FACTORS AND INDICES

FACTORS

The number 2 is a *factor* of 12, since 2 will divide exactly into 12.

The number 12 may be expressed as the product of two factors in several different ways, namely:

$$1 \times 12 \qquad 2 \times 6 \qquad \text{or} \qquad 3 \times 4$$

EXERCISE 12a Express each of the following numbers as the product of two factors, giving all possibilities:

1. 18	**5.** 30	**9.** 48	**13.** 80	**17.** 120
2. 20	**6.** 36	**10.** 60	**14.** 96	**18.** 135
3. 24	**7.** 40	**11.** 64	**15.** 100	**19.** 144
4. 27	**8.** 45	**12.** 72	**16.** 108	**20.** 160

MULTIPLES

12 is a *multiple* of 2 since 12 contains the number 2 a whole number of times.

EXERCISE 12b
1. Write down all the multiples of 3 between 20 and 40.

2. Write down all the multiples of 5 between 19 and 49.

3. Write down all the multiples of 7 between 25 and 60.

4. Write down all the multiples of 11 between 50 and 100.

5. Write down all the multiples of 13 between 25 and 70.

PRIME NUMBERS

Some numbers can be expressed as the product of two factors in only one way. For example, the only factors of 3 are 1 and 3 and the only factors of 5 are 1 and 5. Any number bigger than 1 that is of this type is called a *prime number*. Note that 1 is *not* a prime number.

EXERCISE 12c
1. Which of the following numbers are prime numbers?

   2, 3, 4, 5, 6, 7, 8, 9, 10, 11, 12, 13

2. Write down all the prime numbers between 20 and 30.

3. Write down all the prime numbers between 30 and 50.

4. Which of the following numbers are prime numbers?

5, 10, 19, 29, 39, 49, 61

5. Which of the following numbers are prime numbers?

41, 57, 91, 101, 127

6. Are the following statements true or false?

a) All prime numbers are odd numbers.
b) All odd numbers are prime numbers.
c) All prime numbers between 10 and 100 are odd numbers.
d) The only even prime number is 2.
e) There are six prime numbers less than 10.

INDICES

The accepted shorthand way of writing $2 \times 2 \times 2 \times 2$ is 2^4. We read this as "2 to the power of 4" or "2 to the fourth". The 4 is called the *index*. Hence $16 = 2 \times 2 \times 2 \times 2 = 2^4$ and similarly $3^3 = 3 \times 3 \times 3 = 27$.

EXERCISE 12d Write the following products in index form:

1. $2 \times 2 \times 2$

2. $3 \times 3 \times 3 \times 3$

3. $5 \times 5 \times 5 \times 5$

4. $7 \times 7 \times 7 \times 7 \times 7$

5. $2 \times 2 \times 2 \times 2 \times 2$

6. $3 \times 3 \times 3 \times 3 \times 3 \times 3$

7. $13 \times 13 \times 13$

8. 19×19

9. $2 \times 2 \times 2 \times 2 \times 2 \times 2 \times 2$

10. $6 \times 6 \times 6 \times 6$

Find the value of:

3^5

$$3^5 = 3 \times 3 \times 3 \times 3 \times 3 = 243$$

11. 2^5 **13.** 5^2 **15.** 3^2 **17.** 3^4

12. 3^3 **14.** 2^3 **16.** 7^2 **18.** 2^4

Express the following numbers in index form:

19. 4   **21.** 8   **23.** 49   **25.** 32

20. 9   **22.** 27   **24.** 25   **26.** 64

We can now write any number as the product of prime numbers in index form. Consider the number 108:

$$108 = 12 \times 9$$
$$= 4 \times 3 \times 9$$
$$= 2 \times 2 \times 3 \times 3 \times 3$$

i.e. $$108 = 2^2 \times 3^3$$

Therefore 108 expressed as the product of prime numbers or factors in index form is $2^2 \times 3^3$. Similarly

$$441 = 9 \times 49$$
$$= 3 \times 3 \times 7 \times 7$$
$$= 3^2 \times 7^2$$

EXERCISE 12e Write the following products in index form:

> $2 \times 2 \times 3 \times 3 \times 3$
>
> $2 \times 2 \times 3 \times 3 \times 3 = 2^2 \times 3^3$

1. $2 \times 2 \times 7 \times 7$ **6.** $3 \times 11 \times 11 \times 2 \times 2$

2. $3 \times 3 \times 3 \times 5 \times 5$ **7.** $7 \times 7 \times 7 \times 3 \times 5 \times 7 \times 3$

3. $5 \times 5 \times 5 \times 13 \times 13$ **8.** $13 \times 5 \times 13 \times 5 \times 13$

4. $2 \times 3 \times 3 \times 5 \times 2 \times 5$ **9.** $3 \times 5 \times 5 \times 3 \times 7 \times 3 \times 7$

5. $2 \times 2 \times 3 \times 2 \times 3 \times 5 \times 5$ **10.** $2 \times 3 \times 2 \times 5 \times 3 \times 5$

Find the value of:

> $2^3 \times 3^2$
>
> $2^3 \times 3^2 = 2 \times 2 \times 2 \times 3 \times 3 = 72$

11. $2^2 \times 3^3$ **13.** $2^4 \times 7$ **15.** $2^2 \times 3^2 \times 5$

12. $3^2 \times 5^2$ **14.** $2^2 \times 3^2$ **16.** $2 \times 3^2 \times 7$

FINDING PRIME FACTORS

The following rules may help us to decide whether a given number has certain prime numbers as factors:

A number is divisible

by 2 if the last figure is even

by 3 if the sum of the digits is divisible by 3

by 5 if the last figure is 0 or 5

EXERCISE 12f

> Is 446 divisible by 2?
>
> Since the last figure is even, 446 is divisible by 2.

> Is 1683 divisible by 3?
>
> The sum of the digits is $1+6+8+3 = 18$, which is divisible by 3. Therefore 1683 is divisible by 3.

> Is 7235 divisible by 5?
>
> Since the last digit is 5, 7235 is divisible by 5.

1. Is 525 divisible by 3?

2. Is 747 divisible by 5?

3. Is 2931 divisible by 3?

4. Is 740 divisible by 5?

5. Is 543 divisible by 5?

6. Is 1424 divisible by 2?

7. Is 9471 divisible by 3?

8. Is 2731 divisible by 2?

> Is 8820 divisible by 15?
>
> 8820 is divisible by 5 since it ends in 0.
>
> 8820 is divisible by 3 since $8+8+2 = 18$ which is divisible by 3.
>
> 8820 is therefore divisible by both 5 and 3, i.e. it is divisible by 5×3 or 15.

9. Is 10 752 divisible by 6?

10. Is 21 168 divisible by 6?

11. Is 30 870 divisible by 15?

EXPRESSING A NUMBER IN PRIME FACTORS

To express a number in prime factors start by trying to divide by 2 and keep on until you can no longer divide exactly by 2. Next try 3 in the same way, then 5 and so on for each prime number until you are left with 1.

EXERCISE 12g

Express 720 in prime factors.

(Test for the prime factors in order, 2 first, then 3, and so on.)

2	720
2	360
2	180
2	90
3	45
3	15
5	5
	1

Therefore $720 = 2 \times 2 \times 2 \times 2 \times 3 \times 3 \times 5$

i.e. $720 = 2^4 \times 3^2 \times 5$

Express each of the following numbers in prime factors:

1. 24   **3.** 63   **5.** 136   **7.** 216   **9.** 405

2. 28   **4.** 72   **6.** 84   **8.** 528   **10.** 784

HIGHEST COMMON FACTOR (HCF)

The highest common factor of two or more numbers is the largest number that divides exactly into each of them.

For example 8 is the HCF of 16 and 24

and 15 is the HCF of 45, 60 and 120.

EXERCISE 12h State the HCF of:

1. 9, 12   **5.** 25, 50, 75   **9.** 25, 35, 50, 60

2. 8, 16   **6.** 22, 33, 44   **10.** 36, 44, 52, 56

3. 12, 24   **7.** 21, 42, 84   **11.** 15, 30, 45, 60

4. 14, 42   **8.** 39, 13, 26   **12.** 10, 18, 20, 36

LOWEST COMMON MULTIPLE (LCM)

The lowest common multiple of two or more numbers is the smallest number that divides exactly by each of the numbers. For example the LCM of 8 and 12 is 24 since both 8 and 12 divide exactly into 24. Similarly the LCM of 4, 6 and 9 is 36.

EXERCISE 12i State the LCM of:

1. 3, 5	**5.** 3, 9, 12	**9.** 9, 12, 18
2. 6, 8	**6.** 10, 15, 20	**10.** 18, 27, 36
3. 5, 15	**7.** 12, 16, 24	**11.** 9, 12, 36
4. 9, 12	**8.** 4, 5, 6	**12.** 6, 7, 8

PROBLEMS INVOLVING HCFs AND LCMs

EXERCISE 12j

1. What is the smallest sum of money that can be made up of an exact number of 20 p pieces or of 50 p pieces?

2. Find the least sum of money into which 24 p, 30 p and 54 p will divide exactly.

3. Find the smallest length that can be divided exactly into equal sections of length 5 m or 8 m or 12 m.

4. A room measures 450 cm by 350 cm. Find the side of the largest square tile that can be used to tile the floor without any cutting.

5. Two cars travel around a Scalextric track, the one completing the circuit in 6 seconds and the other in $6\frac{1}{2}$ seconds. If they leave the starting line together how long will it be before they are again side by side?

6. If I go up a flight of stairs two at a time I get to the top without any being left over. If I then try three at a time and again five at a time, I still get to the top without any being left over. Find the shortest flight of stairs for which this is possible. How many would remain if I were able to go up seven at a time?

7. In the first year of a large comprehensive school it is possible to divide the pupils into equal sized classes of either 24 or 30 or 32 and have no pupils left over. Find the size of the smallest entry that makes this possible. How many classes will there be if each class is to have 24 pupils?

13 SETS

SET NOTATION

A *set* is a collection of things that have something in common. We talk about a set of drawing instruments, a set of cutlery and a set of books.

Name some sets.

Things which belong to a set are called *members* or *elements*. These members or elements are usually separated by commas and written down between curly brackets or braces { }.

Instead of writing   "the set of musical instruments"

we write   {musical instruments}

EXERCISE 13a **1.** Use the correct set notation to write down the following sets:

> The set of British cars = {British cars}

    a) The set of foreign cars.
    b) The set of pupils in my class.
    c) The set of subjects I study at school.
    d) The set of furniture in this room.

2. Write down two members from each of the sets given in question 1.

> {British cars}
>
>   Jaguar, Rover

We do not have to list all the members of a set; frequently we can use words to describe the members in a set.

For example instead of {Sunday, Monday,..., Saturday}
we could say {the days of the week}

and instead of {5, 6, 7, 8, 9} we could say
{whole numbers from 5 to 9 inclusive}.

EXERCISE 13b

> Describe in words the set {a, b, c, d, e}.
>
>   {a, b, c, d, e} = {the first five letters of the alphabet}

In questions 1 to 10 describe in words the given sets:

1. {w, x, y, z}

2. {January, June, July}

3. {June, July, August}

4. {England, Scotland, Wales, Northern Ireland}

5. {London, Edinburgh, Cardiff, Belfast}

6. {2, 4, 6, 8, 10, 12}

7. {1, 2, 3, 4, 5, 6}

8. {2, 3, 5, 7, 11, 13}

9. {45, 46, 47, 48, 49, 50}

10. {15, 20, 25, 30, 35}

In questions 11 to 15 describe a set which includes the given members and state another member of it:

11. {Peter, John, David, Richard}

12. {overcoat, raincoat, anorak, trenchcoat}

13. {Rice Crispies, Cornflakes, All Bran, Weetabix}

14. {daffodil, crocus, hyacinth, tulip}

15. {Macbeth, Julius Caesar, King Lear, Romeo and Juliet}

In the remaining questions list the members in the given sets:

{months of the year beginning with the letter M} = {March, May}

16. {whole numbers greater than 10 but less than 16}

17. {the first eight letters of the alphabet}

18. {the letters used in the word "mathematics"}

19. {the countries forming Great Britain}

20. {the countries forming the United Kingdom}

21. {subjects I study}

22. {oceans of the world}

23. {foods I ate for breakfast this morning}

24. {prime numbers less than 20}

25. {even numbers less than 20}

26. {odd numbers between 20 and 30}

27. {multiples of 3 between 10 and 31}

28. {multiples of 7 between 15 and 50}

29. {capital cities in Great Britain}

30. {national television channels}

THE SYMBOL ∈

Instead of writing

"August is a member of the set of months of the year"

we write   August ∈ {months of the year}

The symbol ∈ means "is a member of" or "is an element of".

EXERCISE 13c Write the following statements in set notation:

1. Apple is a member of the set of fruit.

2. Shirt is a member of the set of clothing.

3. Dog is a member of the set of domestic animals.

4. Geography is a member of the set of school subjects.

5. Carpet is a member of the set of floor coverings.

6. Hairdressing is a member of the set of occupations.

THE SYMBOL ∉

We are all aware that August is *not* a member of the set of days of the week.

Since we have chosen ∈ to mean "is a member of" we use ∉ to mean "is *not* a member of". We can therefore write

"August is not a member of the set of days of the week"

as   August ∉ {days of the week}

EXERCISE 13d Write the following statements in set notation:

1. Orange is not a member of the set of animals.

2. Cat is not a member of the set of fruit.

3. Table is not a member of the set of trees.

4. Shirt is not a member of the set of subjects I study.

5. Anne is not a member of the set of boys' names.

6. Chisel is not a member of the set of buildings.

7. Cup is not a member of the set of bedroom furniture.

8. Rover is not a member of the set of foreign cars.

9. Aeroplane is not a member of the set of foreign countries.

10. Curry is not a member of the set of breeds of dogs.

Now write each of the following in set notation:

11. Porridge is a member of the set of breakfast cereals.

12. Electricity is not a member of the set of building materials.

13. Water is not a member of the set of metals.

14. Spider is a member of the set of living things.

15. Saturday is a member of the set of days of the week.

16. A salmon is a fish.

17. August is not the name of a day of the week.

18. Spain is a European country.

19. Brazil is not an Asian country.

Write down the meaning of:

20. Rugby football $\in$ {team games}

21. Shoes $\notin$ {beverages}

22. Hockey $\notin$ {electrical appliances}

23. Needle $\in$ {metal objects}

24. Susan $\notin$ {boys' names}

25. Using the correct notation write down
a) three members that belong to
b) three members that do not belong to

{dairy produce}

26. Using the correct notation, write down

a) three members that belong to

b) three members that do not belong to

{clothes}

FINITE AND INFINITE SETS

Frequently, we need to refer to a set several times. When this is so we label the set with a capital letter. For example

$$A = \{\text{months of the year beginning with the letter J}\}$$

or $\qquad\qquad A = \{\text{January, June, July}\}$

In many cases it is not possible to list all the members of a set. When this is so we write down the first few members followed by dots.

For example if $\qquad N = \{\text{positive whole numbers}\}$

we could write $\qquad\qquad N = \{1, 2, 3, 4, \ldots\}$

Similarly if

$$X = \{\text{even numbers}\} \quad \text{and} \quad Y = \{\text{odd numbers}\}$$

we could write

$$X = \{2, 4, 6, 8, \ldots\} \quad \text{and} \quad Y = \{1, 3, 5, 7, \ldots\}$$

Sets like N, X and Y are called *infinite sets* because there is no limit to the number of members each contains. When we can write down, or count, all the members in a set, the set is called a *finite set*.

EQUAL SETS

When two sets have exactly the same members they are said to be equal.

If $\qquad\qquad A = \{2, 4, 6, 8\} \quad \text{and} \quad B = \{6, 4, 8, 2\}$

then $\qquad\qquad\qquad\qquad A = B$

Similarly, if $\qquad X = \{\text{prime numbers less than 8}\}$
$$= \{2, 3, 5, 7\}$$

and $\quad Y = \{\text{whole numbers up to 7 inclusive except 1, 4 and 6}\}$
$$= \{2, 3, 5, 7\}$$
then $\qquad\qquad\qquad\qquad X = Y$

The order in which the members are listed does not matter, neither does the way in which the sets are described.

EXERCISE 13e Determine whether or not the following sets are equal:

1. $A = \{$chair, table, desk, blackboard$\}$

    $B = \{$desk, blackboard, table, chair$\}$

2. $X = \{$d, i, k, f, w$\}$

    $Y = \{$f, w, k, i$\}$

3. $V = \{4, 6, 8, 10, 12\}$

    $W = \{$even numbers from 4 to 12 inclusive$\}$

4. $C = \{$i, e, a, u, o$\}$

    $D = \{$vowels$\}$

5. $P = \{$Capital cities of the United Kingdom$\}$

    $Q = \{$Belfast, Edinburgh, Cardiff, London$\}$

EMPTY SET

Have you ever seen a woman with three eyes or a man with four legs? We hope not, for neither exists. There are no members in either of these sets. Such a set is called an *empty* or *null* set and is written $\{\quad\}$ or $\emptyset$.

EXERCISE 13f
1. Give some examples of empty sets.

2. Which of the following sets are empty?
    a) $\{$dogs with wings$\}$
    b) $\{$men who have landed on the moon$\}$
    c) $\{$children more than 5 m tall$\}$
    d) $\{$cars that can carry 100 people$\}$
    e) $\{$men more than 100 years old$\}$
    f) $\{$dogs without tails$\}$

SUBSETS

If $A = \{$Paul, Peter, John, Mary, Jane$\}$ and $B = \{$Mary, Jane$\}$ we see that all the members of B are also members of A.
We say that B is a subset of A and write this $B \subset A$.

If $X = \{$a, b, c$\}$ then $\{$a, b, c$\}$, $\{$a, b$\}$, $\{$b, c$\}$, $\{$a, c$\}$, $\{$a$\}$, $\{$b$\}$, $\{$c$\}$ and $\emptyset$ are all subsets of X.

Note that both X and $\emptyset$ are considered to be subsets of X. Subsets that do not contain all the members of X are called *proper subsets*. All the subsets given above except $\{$a, b, c$\}$ are therefore proper subsets of X.

EXERCISE 13g **1.** If $A = \{w, x, y, z\}$ write down all the subsets of A that have two members.

2. If $B = \{$Anne, Bernard, Clive, Doris$\}$ write down all the subsets of B that have two female members.

3. If $N = \{1, 2, 3, \ldots, 10\}$ list the following subsets of N:
$A = \{$odd numbers$\}$
$B = \{$even numbers$\}$
$C = \{$prime numbers$\}$

4. Give a subset with at least three members for each of the following sets:
a) $\{$British cities$\}$
b) $\{$lakes$\}$
c) $\{$oceans$\}$
d) $\{$first division football clubs$\}$

5. If $X = \{1, 2, 3, 5, 7, 11, 13\}$ which of the following sets are proper subsets of X?
a) $\{$odd numbers less than 6$\}$
b) $\{$even numbers less than 4$\}$
c) $\{$prime numbers less than 14$\}$
d) $\{$odd numbers between 10 and 14$\}$

THE UNIVERSAL SET

Consider the set $X = \{$whole numbers less than 16$\}$,
i.e. the set $\{1, 2, 3, 4, \ldots, 15\}$.

Now consider the sets A, B and C whose members are in X such that

$$A = \{\text{prime numbers}\} = \{2, 3, 5, 7, 11, 13\}$$
$$B = \{\text{multiples of 3}\} = \{3, 6, 9, 12, 15\}$$
$$C = \{\text{multiples of 5}\} = \{5, 10, 15\}$$

The original set $\{1, 2, 3, 4, \ldots, 15\}$ is called the *universal set* for the sets A, B and C. It is a set that contains all the members that occur in the sets A, B and C as well as some other members that are not found in any of these three. The *universal set* is denoted by the symbol $\mathscr{E}$ or $\mathscr{U}$.

EXERCISE 13h

Suggest a suitable universal set for $\{$cup, plate, saucer$\}$.
$$\mathscr{E} = \{\text{crockery}\}$$

Suggest a suitable universal set for:

1. {8, 12, 16, 17, 20}

2. {vowels}

3. {Italian cities}

4. {pupils with fair hair}

5. {cats with three legs}

6. {hamsters}

Write down at least two subsets for each of the following universal sets:

$\mathscr{E}$ = {boys' names}

Two subsets are {John, Peter, Paul} and {Dino, Fritz, Alec}.

7. $\mathscr{E}$ = {girls' names}

8. $\mathscr{E}$ = {European countries}

9. $\mathscr{E}$ = {African countries}

10. $\mathscr{E}$ = {Members of Parliament}

11. $\mathscr{E}$ = {school subjects}

12. $\mathscr{E}$ = {colours}

Suggest a universal set for:

13. {squares, rectangles, rhombuses}

14. {houses, flats, bungalows}

15. {Rover, Jaguar, Metro, Maestro}

16. {trainers, shoes, sandals, boots}

17. {golfers, football players, skiers, sprinters}

VENN DIAGRAMS

Many years ago a Cambridge mathematician named John Venn studied the algebra of sets and introduced the diagrams which now bear his name. In a Venn diagram the universal set ($\mathscr{E}$) is usually represented by a rectangle and subsets of the universal set are usually shown as circles inside the rectangle. There is nothing special about circles — any convenient shape would do.

If $\mathscr{E}$ = {school children} then A could be {pupils in my school}, i.e. A is a subset of $\mathscr{E}$.

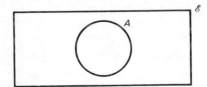

Similarly, if $B = \{$pupils in the next school to my school$\}$ the diagram would be

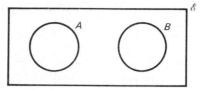

Two sets like these, which have no common members, are called

disjoint sets

If $C = \{$pupils in my class$\}$ then, because all the members of C are also members of A, i.e. C is a proper subset of A, the Venn diagram is

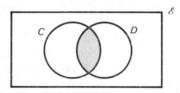

If $D = \{$my school friends$\}$ the corresponding Venn diagram could be

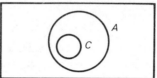

The shaded region shows the friends I have who are in my class. These friends belong to both sets. The unshaded region of D represents friends I have in school who are not in my class.

UNION OF TWO SETS

In my class

$A = \{$pupils good at maths$\} = \{$Frank, Javed, Asif, Sian$\}$

and

$B = \{$pupils good at French$\} = \{$Bina, Asif, Polly, Frank$\}$

If the universal set is $\{$all the pupils in my class$\}$ the names could be placed in a Venn diagram as follows:

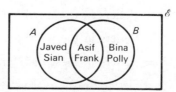

If we write down the set of all the members of my class who are good at *either* maths *or* French we have the set {Javed, Sian, Asif, Frank, Bina, Polly}. This is called the *union* of the sets A and B and is denoted by

$$A \cup B$$

Similarly if $X = \{1, 2, 3, 5\}$ and $Y = \{2, 4, 6\}$ we can illustrate these sets in the following Venn diagram:

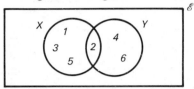

and write the union of the two sets X and Y

$$X \cup Y = \{1, 2, 3, 4, 5, 6\}$$

To find the union of two sets, write down all the members of the first set, then all the members of the second set which have not already been included.

EXERCISE 13i Find the union of the two given sets in each of the following:

> $A = \{3, 6, 9, 12\}$ $B = \{4, 6, 8, 10, 12\}$
>
> $A \cup B = \{3, 4, 6, 8, 9, 10, 12\}$

1. $A = \{$Peter, James, John$\}$ $B = \{$John, Andrew, Paul$\}$

2. $X = \{3, 6, 9, 12\}$ $Y = \{4, 8, 12, 16\}$

3. $P = \{$a, e, i, o, u$\}$ $Q = \{$a, b, c, d, e$\}$

4. $A = \{$a, b, c$\}$ $B = \{$x, y, z$\}$

5. $A = \{$p, q, r, s, t$\}$ $B = \{$p, r, t$\}$

6. $X = \{2, 3, 5, 7\}$ $Y = \{1, 3, 5, 7\}$

7. $X = \{5, 7, 11, 13\}$ $Y = \{6, 8, 10, 12\}$

8. $P = \{$whole numbers that divide exactly into 12$\}$

    $Q = \{$whole numbers that divide exactly into 10$\}$

9. $A = \{$letters in the word "classroom"$\}$

    $B = \{$letters in the word "school"$\}$

10. $P = \{$letters in the word "arithmetic"$\}$

    $Q = \{$letters in the word "algebra"$\}$

To represent the union of two sets in a Venn diagram we shade the combined area representing the two sets. This shaded area may occur in three ways.

a) When sets *A* and *B* have some common members.

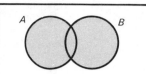

b) When *A* and *B* have no common member, i.e. when they are disjoint.

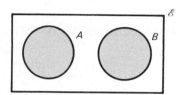

c) When *B* is a proper subset of *A*.

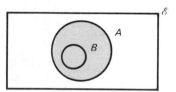

EXERCISE 13j Draw suitable Venn diagrams to show the unions of the following sets:

$$P = \{3, 6, 9, 12, 15\} \qquad Q = \{3, 5, 7, 9, 11, 15\}$$

$$P \cup Q = \{3, 5, 6, 7, 9, 11, 12, 15\}$$

1. $A = \{p, q, r, s\}$ $B = \{r, s, t, u\}$

2. $X = \{1, 3, 5, 7, 9\}$ $Y = \{2, 4, 6, 8, 10\}$

3. $P = \{a, b, c, d, e, f, g\}$ $Q = \{c, d, g\}$

4. $E = \{\text{triangles}\}$ $F = \{\text{isosceles triangles}\}$

5. $G = \{\text{even numbers}\}$ $H = \{\text{odd numbers}\}$

6. $M = \{\text{acute angles}\}$ $N = \{\text{obtuse angles}\}$

7. $A = \{3, 6, 9, 12, 15\}$ $B = \{4, 6, 8, 10, 12, 14\}$

8. $P = \{$letters in the word "Donald"$\}$

$Q = \{$letters in the word "London"$\}$

9. $X = \{$Marc, Leslie, Joe, Claude$\}$

$Y = \{$Leslie, Sita, Joe, Yvette$\}$

10. $A = \{$letters in the word "metric"$\}$

$B = \{$letters in the word "imperial"$\}$

INTERSECTION OF SETS

If we return to the set of pupils in my class

$A = \{$pupils good at maths$\}$
$\quad = \{$Frank, Javed, Asif, Sian$\}$

and

$B = \{$pupils good at French$\}$
$\quad = \{$Bina, Asif, Polly, Frank$\}$

then Frank and Asif form the set of pupils who are good at *both* maths and French. The members that are in both sets give what is called the *intersection* of the sets A and B.

The intersection of two sets A and B is written $A \cap B$, i.e. for the given sets

$$A \cap B = \{\text{Frank, Asif}\}$$

EXERCISE 13k Find the intersection of the following pairs of sets:

> $X = \{1, 2, 3, 4, 5, 6\}$ $Y = \{1, 2, 3, 5, 7\}$
>
> $X \cap Y = \{1, 2, 3, 5\}$

1. $A = \{3, 6, 9, 12\}$ $B = \{5, 6, 7, 8, 9\}$

2. $X = \{4, 8, 12, 16, 20\}$ $Y = \{4, 12, 20\}$

3. $P = \{$Bob, Ken, Colin, Alice$\}$

$Q = \{$Alice, Bill, Hans, Bob$\}$

4. $C = \{$o, p, q, r, s, t$\}$ $D = \{$a, e, i, o, u$\}$

5. $A = \{$tomato, cabbage, apple, pear$\}$

$B = \{$cabbage, tomato$\}$

6. $M = \{$prime numbers less than 12$\}$

$N = \{$odd numbers less than 12$\}$

7. $P = \{4, 8, 12, 16\}$ $Q = \{8, 16, 24, 48\}$

8. $A = \{$factors of 12$\}$ $B = \{$factors of 10$\}$

9. $X = \{$letters in the word "twice"$\}$

     $Y = \{$letters in the word "sweat"$\}$

10. $P = \{$letters in the word "metric"$\}$

     $Q = \{$letters in the word "imperial"$\}$

EXERCISE 13I Draw suitable Venn diagrams to show the intersections of the following sets:

$X = \{1, 2, 3, 4, 5, 6\}$ $Y = \{2, 3, 5, 7, 11\}$

$X \cap Y = \{2, 3, 5\}$

1. $A = \{1, 3, 5, 7, 9, 11\}$ $B = \{2, 3, 4, 5, 6, 7\}$

2. $P = \{$John, David, Dino, Kay$\}$

    $Q = \{$Pete, Dino, Omar, John$\}$

3. $X = \{$a, e, i, o, u$\}$ $Y = \{$b, f, o, w, u$\}$

4. $A = \{$oak, ash, elm, pine$\}$

    $B = \{$teak, oak, sapele, elm$\}$

5. $X = \{$poodle, greyhound, boxer$\}$

    $Y = \{$pug, collie, boxer, cairn$\}$

6. $P = \{4, 8, 12, 16\}$ $Q = \{8, 16, 24, 48\}$

7. $A = \{$factors of 12$\}$ $B = \{$factors of 20$\}$

8. $X = \{$letters in the word "think"$\}$

    $Y = \{$letters in the word "flint"$\}$

9. $A = \{$letters in the word "arithmetic"$\}$

    $B = \{$letters in the word "geometry"$\}$

10. $P = \{$prime numbers less than 10$\}$

     $Q = \{$odd numbers less than 15$\}$

14 AREA

COUNTING SQUARES

The area of a shape or figure is the amount of surface enclosed within the lines which bound it. Below, six letters have been drawn on squared paper.

We can see, by counting squares, that the area of the letter E is 15 squares.

EXERCISE 14a What is the area of:

1. The letter T?

2. The letter H?

Sometimes the squares do not fit exactly on the area we are finding. When this is so we count a square if more than half of it is within the area we are finding, but exclude it if more than half of it is outside.

By counting squares in this way the approximate area of the letter A is 13 squares.

What is the approximate area of:

3.   The letter P?

4.   The letter O?

The next set of diagrams shows the outlines of three leaves.

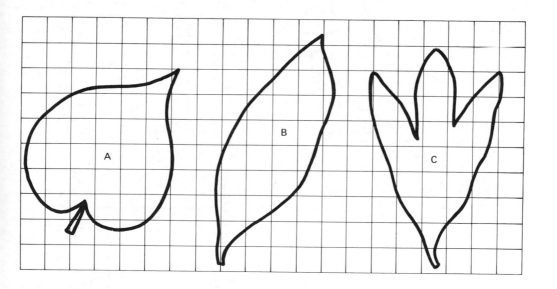

By counting squares find the approximate area of:

5.   The leaf outline marked A.

6.   The leaf outline marked B.

7.   The leaf outline marked C.

8.   Which leaf has

a) the largest area   b) the smallest area?

In each of the following questions find the area of the given figure by counting squares:

9.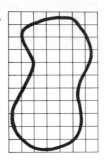

13.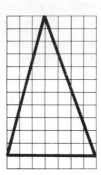

10.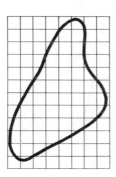

14.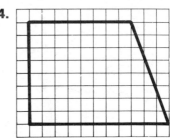

11.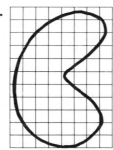

15.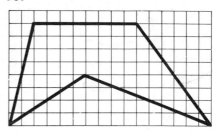

12.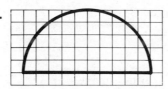

16.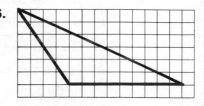

UNITS OF AREA

There is nothing special about the size of square we have used. If other people are going to understand what we are talking about when we say that the area of a certain shape is 12 squares, we must have a square or unit of area which everybody understands and which is always the same.

A metre is a standard length and a square with sides 1 m long is said to have an area of one square metre. We write one square metre as $1 m^2$. Other agreed lengths such as millimetres, centimetres and kilometres, are also in use. The unit of area used depends on what we are measuring.

We could measure the area of a small coin in square millimetres (mm^2), the area of the page of a book in square centimetres (cm^2), the area of a roof in square metres (m^2) and the area of a county in square kilometres (km^2).

AREA OF A SQUARE

The square is the simplest figure of which to find the area. If we have a square whose side is 4 cm long it is easy to see that we must have 16 squares, each of side 1 cm, to cover the given square,

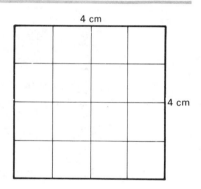

i.e. the area of a square of side 4 cm is $16 cm^2$

AREA OF A RECTANGLE

If we have a rectangle measuring 6 cm by 4 cm we require 4 rows each containing 6 squares of side 1 cm to cover this rectangle,

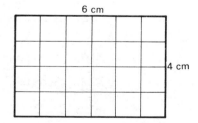

i.e. the area of the rectangle $= 6 \times 4 cm^2$

$$= 24 cm^2$$

A similar result can then be found for a rectangle of any size; for example a rectangle of length 4 cm and breadth $2\frac{1}{2}$ cm has an area of $4 \times 2\frac{1}{2}$ cm^2.

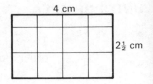

In general, for any rectangle

$$\text{Area} = \text{length} \times \text{breadth}$$

EXERCISE 14b Find the area of each of the following shapes, clearly stating the units involved:

1. A square of side 2 cm

2. A square of side 8 cm

3. A square of side 10 cm

4. A square of side 5 cm

5. A square of side 1.5 cm

6. A square of side 2.5 cm

7. A square of side 0.7 m

8. A square of side 1.2 cm

9. A square of side $\frac{1}{2}$ km

10. A square of side $\frac{3}{4}$ m

11. A rectangle measuring 5 cm by 6 cm

12. A rectangle measuring 6 cm by 8 cm

13. A rectangle measuring 3 m by 9 m

14. A rectangle measuring 14 cm by 20 cm

15. A rectangle measuring 1.8 mm by 2.2 mm

16. A rectangle measuring 35 km by 42 km

17. A rectangle measuring 1.5 m by 1.9 m

18. A rectangle measuring 4.8 cm by 6.3 cm

19. A rectangle measuring 95 cm by 240 cm

20. A rectangle measuring 150 mm by 240 mm

COMPOUND FIGURES

EXERCISE 14c Frequently it is possible to find the area of a figure by dividing it into two or more rectangles.

Find the areas of the following figures by dividing them into rectangles:

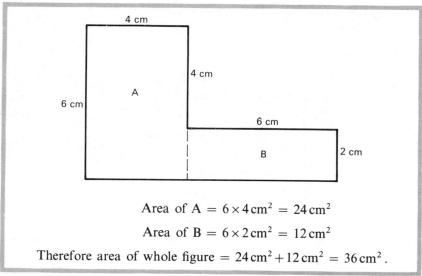

$$\text{Area of A} = 6 \times 4 \, \text{cm}^2 = 24 \, \text{cm}^2$$

$$\text{Area of B} = 6 \times 2 \, \text{cm}^2 = 12 \, \text{cm}^2$$

Therefore area of whole figure $= 24 \, \text{cm}^2 + 12 \, \text{cm}^2 = 36 \, \text{cm}^2$.

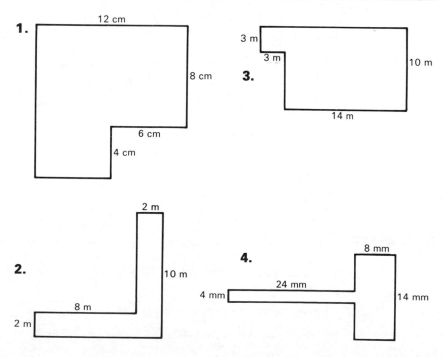

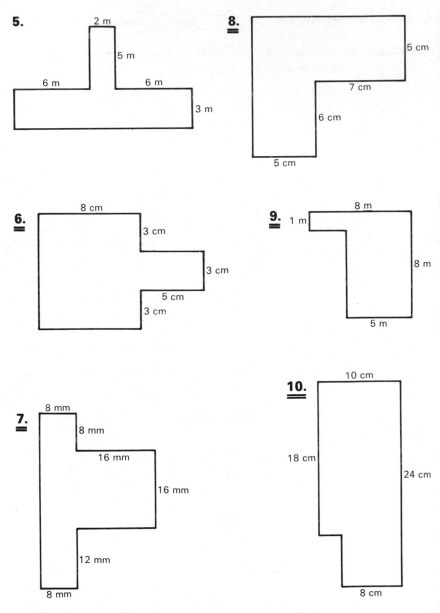

PERIMETER

The total length of the line or lines which bound a figure is called the perimeter.

The perimeter of the square on p. 197 is

$$4\,cm + 4\,cm + 4\,cm + 4\,cm = 16\,cm$$

EXERCISE 14d Find the perimeter of each shape given in Exercise 14b, clearly indicating units.

If we are given a rectangle whose perimeter is 22 cm and told that the length of the rectangle is 6 cm it is possible to find its breadth and its area.

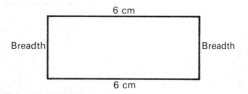

6 cm

Breadth | Breadth

6 cm

The two lengths add up to 12 cm
so the two breadths add up to $(22-12)$ cm $= 10$ cm.

Therefore the breadth is 5 cm.

The area of this rectangle $= 6 \times 5$ cm^2

$$= 30 \text{ cm}^2$$

EXERCISE 14e The following table gives some of the measurements for various rectangles. Fill in the values that are missing:

	Length	Breadth	Perimeter	Area
1.	4 cm		12 cm	
2.	5 cm		14 cm	
3.		3 m	16 m	
4.		6 mm	30 mm	
5.	6 cm			30 cm^2
6.	10 m			120 m^2
7.		4 km		36 km^2
8.		7 mm		63 mm^2
9.		5 cm	60 cm	
10.	21 cm			1680 cm^2

PROBLEMS

EXERCISE 14f Find for each of the following figures

a) the perimeter   b) the area.

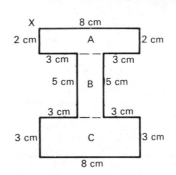

a) Starting at X, the distance all round the figure and back to X is

$8+2+3+5+3+3+8+3+3+5+3+2$ cm $= 48$ cm.

Therefore the perimeter is 48 cm.

b) Divide the figure into three rectangles A, B and C.

Then   the area of A $= 8 \times 2$ cm^2 $= 16$ cm^2

the area of B $= 5 \times 2$ cm^2 $= 10$ cm^2

and   the area of C $= 8 \times 3$ cm^2 $= 24$ cm^2

Therefore the total area $= (16+10+24)$ cm^2 $= 50$ cm^2

1.

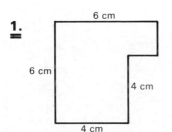

2.

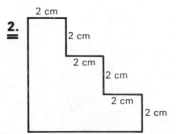

3.

4.

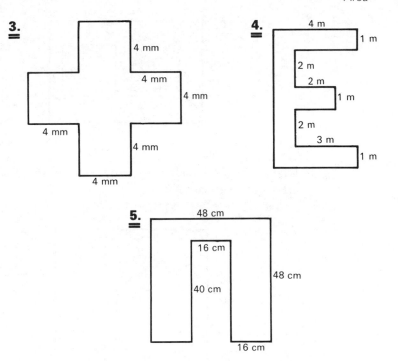

5.

In each of the following figures find the area that is shaded:

6.

8.

7.

9.

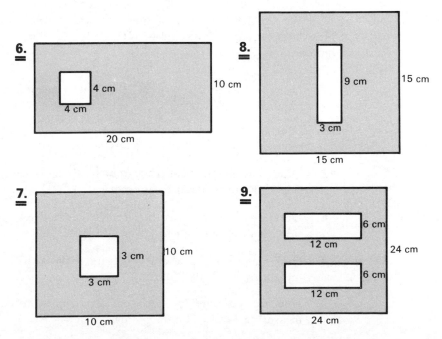

10.

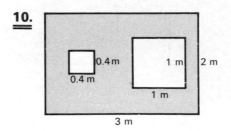

Draw a square of side 6 cm. How many squares of side 2 cm are required to cover it?

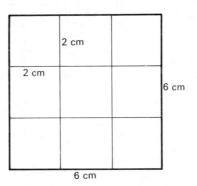

We see that 9 squares of side 2 cm are required to cover the larger square whose side is 6 cm.

1. Draw a square of side 4 cm. How many squares of side 2 cm are required to cover it?

2. Draw a square of side 9 cm. How many squares of side 3 cm are required to cover it?

3. Draw a rectangle measuring 6 cm by 4 cm. How many squares of side 2 cm are required to cover it?

4. Draw a rectangle measuring 9 cm by 6 cm. How many squares of side 3 cm are required to cover it?

5. How many squares of side 5 cm are required to cover a rectangle measuring 45 cm by 25 cm?

6. How many squares of side 4 cm are required to cover a rectangle measuring 1 m by 80 cm?

CHANGING UNITS OF AREA

A square of side 1 cm may be divided into 100 equal squares of side 1 mm,

$$\text{i.e. } 1\,cm^2 = 100\,mm^2$$

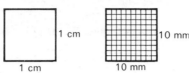

Similarly since 1 m = 100 cm

$$1 \text{ square metre} = 100 \times 100 \text{ square centimetres}$$
$$\text{i.e. } 1\,m^2 = 10\,000\,cm^2$$

and as 1 km = 1000 m

$$1\,km^2 = 1000 \times 1000\,m^2$$
$$\text{i.e. } 1\,km^2 = 1\,000\,000\,m^2$$

When we convert from a unit of area which is large to a unit of area which is smaller we must remember that the number of units will be bigger,

$$\text{e.g. } 2\,km^2 = 2 \times 1\,000\,000\,m^2$$
$$= 2\,000\,000\,m^2$$

and

$$12\,m^2 = 12 \times 10\,000\,cm^2$$
$$= 120\,000\,cm^2$$

while if we convert from a unit of area which is small into one which is larger the number of units will be smaller,

$$\text{e.g. } 500\,mm^2 = \frac{500}{100}\,cm^2$$
$$= 5\,cm^2$$

EXERCISE 14h

Express 5 m² in   a) cm²   b) mm².

a)   Since $1\,m^2 = 100 \times 100\,cm^2$

$$5\,m^2 = 5 \times 100 \times 100\,cm^2$$
$$= 50\,000\,cm^2$$

b)   Since $1\,cm^2 = 100\,mm^2$

$$50\,000\,cm^2 = 50\,000 \times 100\,mm^2$$

Therefore $5\,m^2 = 50\,000\,cm^2 = 5\,000\,000\,mm^2$.

1. Express in cm^2:

   a) 3 m^2 b) 12 m^2 c) 7.5 m^2 d) 82 m^2 e) 8$\frac{1}{2}$ m^2.

2. Express in mm^2:

   a) 14 cm^2 b) 3 cm^2 c) 7.5 cm^2 d) 26 cm^2 e) 32$\frac{1}{2}$ cm^2.

3. Express 0.056 m^2 in   a) cm^2 b) mm^2.

Express 354 000 000 mm^2 in   a) cm^2 b) m^2.

a)   Since   100 mm^2 = 1 cm^2

$$354\,000\,000\,\text{mm}^2 = \frac{354\,000\,000}{100}\,\text{cm}^2$$

$$= 3\,540\,000\,\text{cm}^2$$

b)   Since   100 × 100 cm^2 = 1 m^2

$$3\,540\,000\,\text{cm}^2 = \frac{3\,540\,000}{100 \times 100}\,\text{m}^2$$

$$= 354\,\text{m}^2$$

Therefore 354 000 000 mm^2 = 3 540 000 cm^2 = 354 m^2.

---

4. Express in cm^2:

   a) 400 mm^2 b) 2500 mm^2 c) 50 mm^2
   d) 25 mm^2 e) 734 mm^2

5. Express in m^2:

   a) 5500 cm^2 b) 140 000 cm^2 c) 760 cm^2
   d) 18 600 cm^2 e) 29 700 000 cm^2.

6. Express in km^2:

   a) 7 500 000 m^2 b) 430 000 m^2 c) 50 000 m^2
   d) 245 000 m^2 e) 176 000 000 m^2.

Many questions ask us to find the area of a rectangle but give the length and breadth in different units. When this is so we must change the units so that all the measurements are in the same units.

EXERCISE 14i

Find the area of a rectangle measuring $\frac{1}{2}$ m by 35 cm. Give your answer in cm².

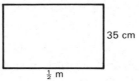

35 cm

$\frac{1}{2}$ m

(Since the answer is to be given in cm² we express both the length and breadth in cm.)

Length of rectangle $= \frac{1}{2}$ m $= \frac{1}{2} \times 100$ cm $= 50$ cm

Therefore   area of rectangle $= 50 \times 35$ cm²

$= 1750$ cm²

Find the area of each of the following rectangles, giving your answer in the unit in brackets:

	Length	Breadth	
1.	10 m	50 cm	(cm²)
2.	6 cm	30 mm	(mm²)
3.	50 m	35 cm	(cm²)
4.	140 cm	1 m	(cm²)
5.	400 cm	200 cm	(m²)
6.	3 m	$\frac{1}{2}$ m	(cm²)
7.	$2\frac{1}{2}$ m	$1\frac{1}{2}$ m	(cm²)
8.	1.5 cm	1.2 cm	(mm²)
9.	0.4 km	0.3 km	(m²)
10.	0.45 km	0.05 km	(m²)

MIXED PROBLEMS

EXERCISE 14j In questions 1 to 4 find   a) the area of the playing surface

b) the perimeter of the playing surface:

1. A soccer field measuring 110 m by 75 m.

2. A rugby pitch measuring 100 m by 70 m.

3. A lacrosse pitch measuring 120 m by 70 m.

4. A tennis court measuring 26 m by 12 m.

5. A roll of wallpaper is 10 m long and 50 cm wide. Find its area in square metres.

6. A school hall measuring 20 m by 15 m is to be covered with square floor tiles of side 50 cm. How many tiles are required?

7. A rectangular carpet measures 4 m by 3 m. Find its area. How much would it cost to clean at 75 p per square metre?

8. The top of my desk is 150 cm long and 60 cm wide. Find its area.

9. How many square linen serviettes, of side 50 cm, may be cut from a roll of linen 25 m long and 1 m wide?

10. How many square concrete paving slabs, each of side $\frac{3}{4}$ m, are required to pave a rectangular yard measuring 9 m by 6 m?

15 PARALLEL LINES AND ANGLES

PARALLEL LINES

Two straight lines that are always the same distance apart, however far they are drawn, are called parallel lines.

The lines in your exercise books are parallel. You can probably find many other examples of parallel lines.

EXERCISE 15a **1.** Using the lines in your exercise book, draw three lines that are parallel. Do not make them all the same distance apart. For example

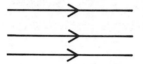

(We use arrows to mark lines that are parallel.)

2. Using the lines in your exercise book, draw two parallel lines. Make them fairly far apart. Now draw a slanting line across them. For example

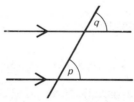

Mark the angles in your drawing that are in the same position as those in the diagram. Are they acute or obtuse angles? Measure your angles marked p and q.

3. Draw a grid of parallel lines like the diagram below. Use the lines in your book for one set of parallels and use the two sides of your ruler to draw the slanting parallels.

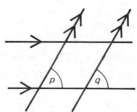

Mark your drawing like the diagram. Are your angles p and q acute or obtuse? Measure your angles p and q.

4. Repeat question 3 but change the direction of your slanting lines.

5. Draw three slanting parallel lines like the diagram below with a horizontal line cutting them. Use the two sides of your ruler and move it along to draw the third parallel line.

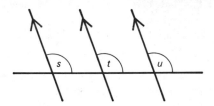

Mark your drawing like the diagram. Decide whether angles *s*, *t* and *u* are acute or obtuse and then measure them.

6. Repeat question 5 but change the slope of your slanting lines.

CORRESPONDING ANGLES

In the exercise above, lines were drawn that crossed a set of parallel lines.

> A line that crosses a set of parallel lines is called a *transversal*.

When you have drawn several parallel lines you should notice that

> two parallel lines on the same flat surface will never meet however far they are drawn.

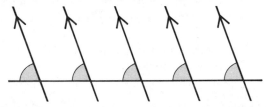

If you draw the diagram above by moving your ruler along you can see that all the shaded angles are equal. These angles are all in corresponding positions: they are all above the transversal and to the left of the parallel lines. Angles like these are called *corresponding angles*.

> When two parallel lines are cut by a transversal, the corresponding angles are equal.

EXERCISE 15b In the diagrams below write down the angle that corresponds to the shaded angle:

1.

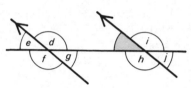

6.

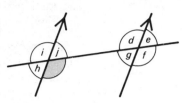

2.

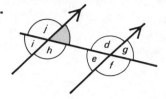

7.

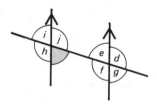

8.

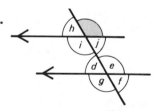

3.

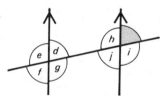

9.

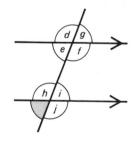

4.

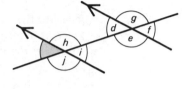

10.

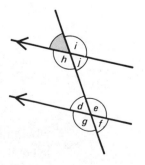

5.

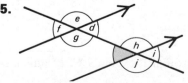

DRAWING PARALLEL LINES (USING A PROTRACTOR)

The fact that the corresponding angles are equal gives us a method for drawing parallel lines.

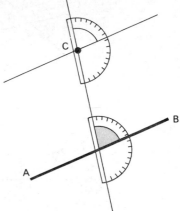

If you need to draw a line through the point C that is parallel to the line AB, first draw a line through C to cut AB.

Use your protractor to measure the shaded angle. Place your protractor at C as shown in the diagram. Make an angle at C the same size as the shaded angle and in the corresponding position.

You can now extend the arm of your angle both ways, to give the parallel line.

EXERCISE 15c **1.** Using your protractor draw a grid of parallel lines like the one in the diagram. (It does not have to be an exact copy.)

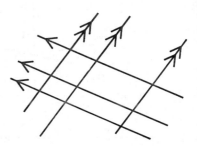

2. Trace the diagram below.

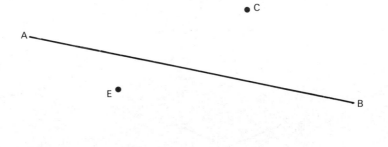

Now draw lines through the points C, D and E so that each line is parallel to AB.

3. Draw a sloping line on your exercise book. Mark a point C above the line. Use your protractor to draw a line through C parallel to your first line.

4. Trace the diagram below.

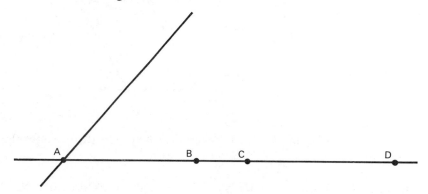

Measure the acute angle at A. Draw the corresponding angles at B, C and D. Extend the arms of your angles so that you have a set of four parallel lines.

In questions 5 to 8 remember to draw a rough sketch before doing the accurate drawing.

5. Draw an equilateral triangle with sides each 8 cm long. Label the corners A, B and C. Draw a line through C that is parallel to the side AB.

6. Draw an isosceles triangle ABC with base AB which is 10 cm long and base angles at A and B which are each 30°. Draw a line through C which is parallel to AB.

7. Draw the triangle as given in question 5 again and this time draw a line through A which is parallel to the side BC.

8. Make an accurate drawing of the figure below where the side AB is 7 cm, the side AD is 4 cm and $\widehat{A} = 60°$.
   (A figure like this is called a *parallelogram*.)

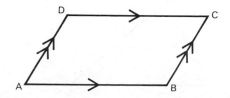

PROBLEMS INVOLVING CORRESPONDING ANGLES

The simplest diagram for a pair of corresponding angles is an F shape.

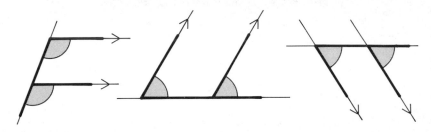

Looking for an F shape may help you to recognise the corresponding angles.

EXERCISE 15d Write down the size of the angle marked *d* in each of the following diagrams:

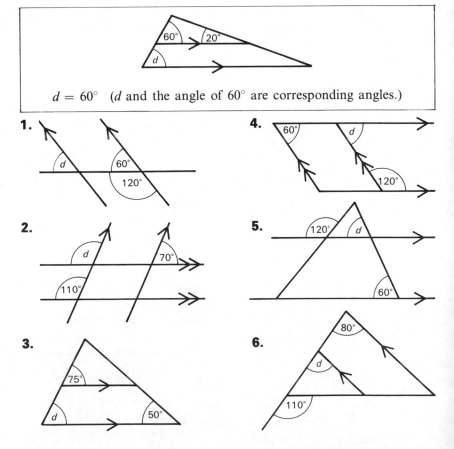

$d = 60°$ (*d* and the angle of 60° are corresponding angles.)

1.

2.

3.

4.

5.

6.

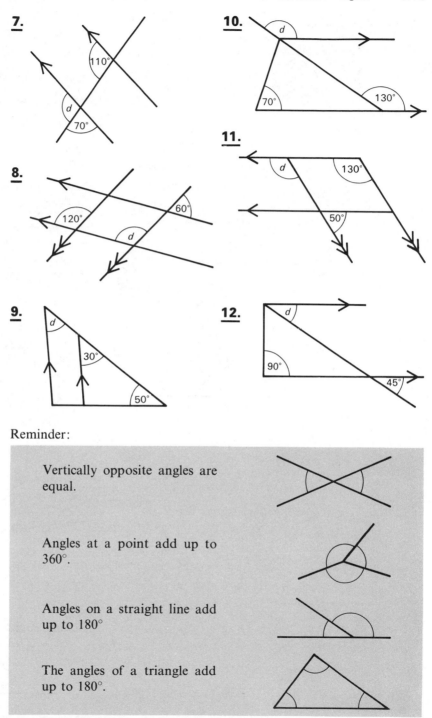

7.

8.

9.

10.

11.

12.

Reminder:

Vertically opposite angles are equal.

Angles at a point add up to 360°.

Angles on a straight line add up to 180°

The angles of a triangle add up to 180°.

You will need these facts in the next exercise.

EXERCISE 15e Find the size of each marked angle:

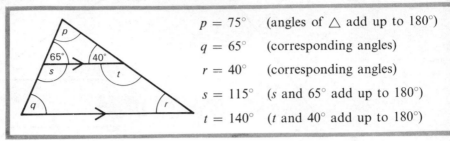

$p = 75°$ (angles of △ add up to 180°)
$q = 65°$ (corresponding angles)
$r = 40°$ (corresponding angles)
$s = 115°$ (s and 65° add up to 180°)
$t = 140°$ (t and 40° add up to 180°)

1.

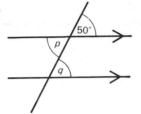

5.

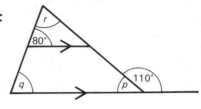

2.

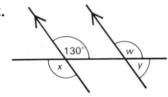

6.

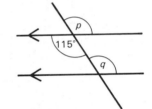

3.

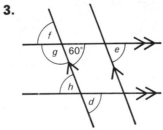

7.

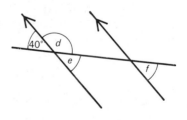

4.

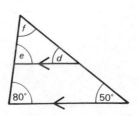

8.

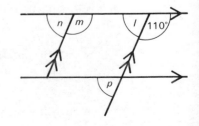

9.

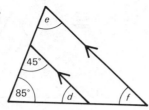

13.

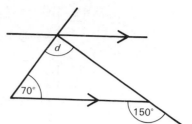

10.

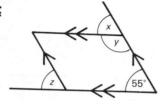

14.

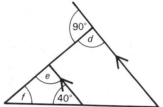

11.

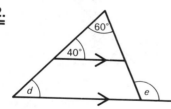

15.

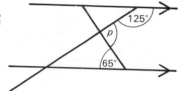

12.

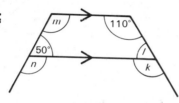

16.

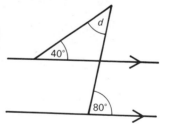

Find the size of angle *d* in questions 17 to 24:

17.

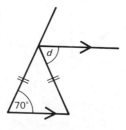

18.

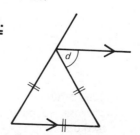

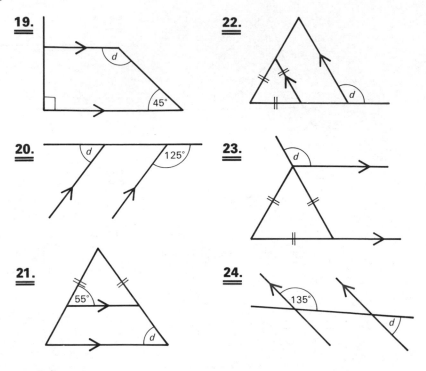

19. *d*, 45°

20. *d*, 125°

21. 55°, *d*

22. *d*

23. *d*

24. 135°, *d*

ALTERNATE ANGLES

Draw a large letter Z. Use the lines of your exercise book to make sure that the outer arms of the Z are parallel.

This letter has rotational symmetry about the point marked with a cross. This means that the two shaded angles are equal. Measure them to make sure.

Draw a large N and make sure that the outer arms are parallel.

This letter also has rotational symmetry about the point marked with a cross, so once again the shaded angles are equal. Measure them to make sure.

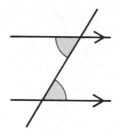

The pairs of shaded angles like those in the Z and N are between the parallel lines and on alternate sides of the transversal.
Angles like these are called *alternate angles*.

When two parallel lines are cut by a transversal, the alternate angles are equal.

The simplest diagram for a pair of alternate angles is a Z shape.

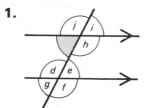
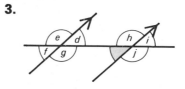

Looking for a Z shape may help you to recognise the alternate angles.

EXERCISE 15f Write down the angle which is alternate to the shaded angle in the following diagrams:

1.

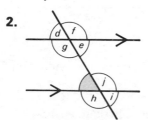

3.

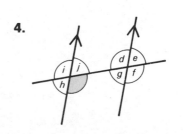

2.

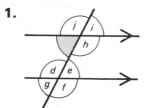

4.

5.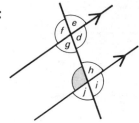

8.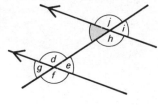

6.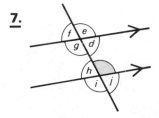

9.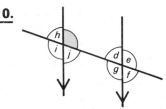

7.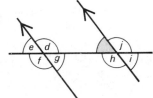

10.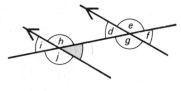

PROBLEMS INVOLVING ALTERNATE ANGLES

Without doing any measuring we can show that alternate angles are equal by using the facts that we already know:

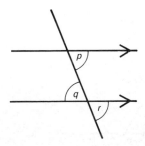

$p = r$ because they are corresponding angles

$q = r$ because they are vertically opposite angles

$\therefore$ $p = q$ and these are alternate angles

EXERCISE 15g Find the size of each marked angle:

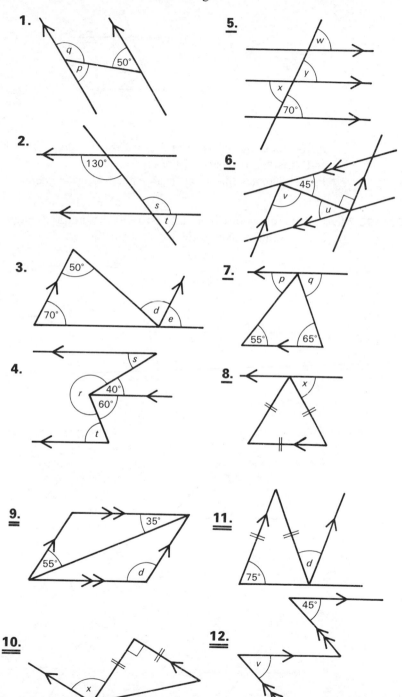

INTERIOR ANGLES

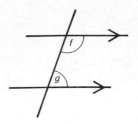

In the diagram above, *f* and *g* are on the same side of the transversal and "inside" the parallel lines.

Pairs of angles like *f* and *g* are called *interior angles*.

EXERCISE 15h In the following diagrams, two of the marked angles are a pair of interior angles. Name them:

1.

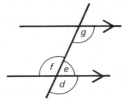

4.

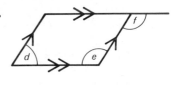

2.

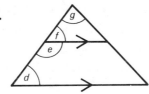

5.

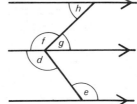

3.

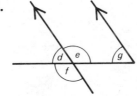

6.

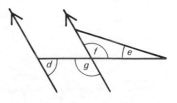

In the following diagrams, use the information given to find the size of p and of q. Then find the sum of p and q:

7.

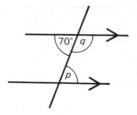

9.

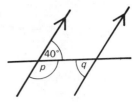

8.

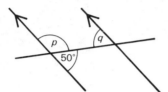

10.

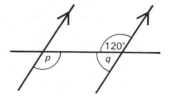

11. Make a large copy of the diagram below. Use the lines of your book to make sure that the outer arms of the "U" are parallel.

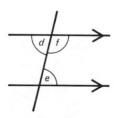

Measure each of the interior angles p and q. Add them together.

> The sum of a pair of interior angles is 180°.

You will probably have realised this fact by now. We can show that it is true from the following diagram.

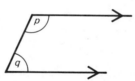

$d+f = 180°$	because they are angles on a straight line
$d = e$	because they are alternate angles
So $e+f = 180°$	

The simplest diagram for a pair of interior angles is a U shape.

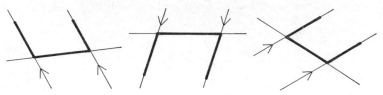

Looking for a U shape may help you to recognise a pair of interior angles.

EXERCISE 15i   Find the size of each marked angle:

1.

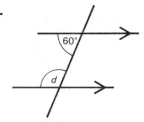

2.

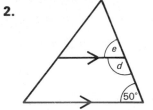

3.

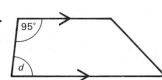

4.

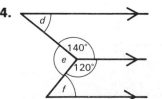

5.

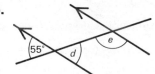

6.

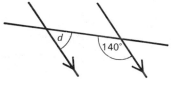

7.

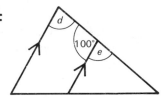

8.

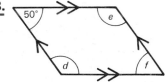

9.

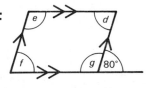

10.

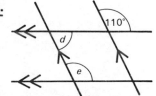

MIXED EXERCISES

You now know that when a transversal cuts a pair of parallel lines

the corresponding (F) angles are equal

the alternate (Z) angles are equal

the interior (U) angles add up to 180°

You can use any of these facts, together with the other angle facts you know, to answer the questions in the following exercises.

EXERCISE 15j Find the size of each marked angle:

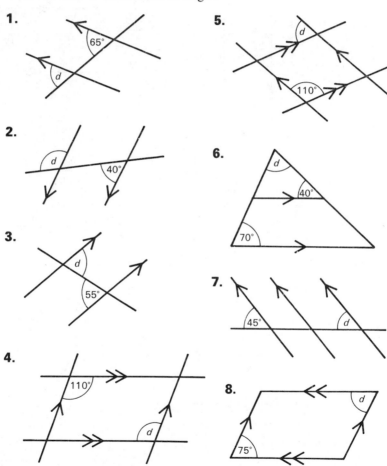

1.

65°

d

2.

d

40°

3.

d

55°

4.

110°

d

5.

d

110°

6.

d

40°

70°

7.

45°

d

8.

d

75°

9. Construct a triangle ABC in which AB = 12 cm, BC = 8 cm and AC = 10 cm. Find the midpoint of AB and mark it D. Find the midpoint of AC and mark it E. Join ED. Measure AD̂E and AB̂C. What can you say about the lines DE and BC?

EXERCISE 15k Find the size of each marked angle:

1.

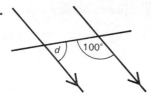

2.

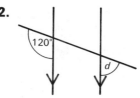

3.

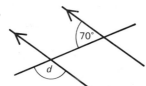

4.

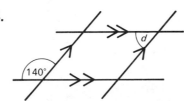

5.

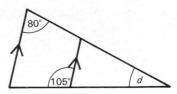

6.

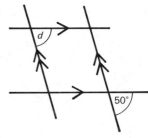

7.

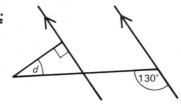

8.

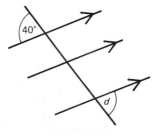

9. Construct the parallelogram on the right, making it full size.

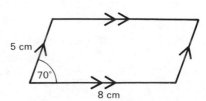

EXERCISE 15I Find the size of each marked angle:

1.

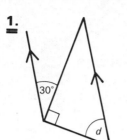

2.

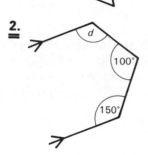

3.

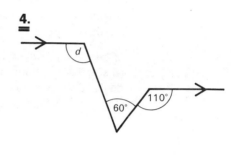

4.

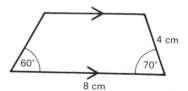

5. Construct an equilateral triangle with sides 7 cm long and label it ABC. Using BC as one side construct another equilateral triangle BCD. Your drawing should look like the diagram on the right. Mark any lines that you think are parallel and give reasons for your decision.

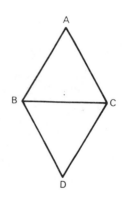

6. Make a full size construction of the trapezium on the right.

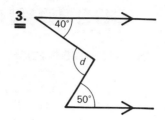

16 COORDINATES

PLOTTING POINTS USING POSITIVE COORDINATES

There are many occasions when you need to describe the position of an object. For example, telling a friend how to find your house, finding a square in the game of battleships, describing the position of an aeroplane showing up on a radar screen. In mathematics we need a quick way to describe the position of a point.

We do this by using squared paper and marking a point O at the corner of one square. We then draw a line through O across the page. This line is called O*x*. Next we draw a line through O up the page. This line is called O*y*. Starting from O we then mark numbered scales on each line.

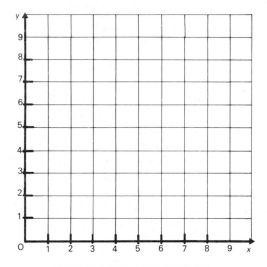

O is called the origin

O*x* is called the *x*-axis

O*y* is called the *y*-axis

We can now describe the position of a point A as follows:
start from O and move 3 squares along O*x*,
then move 5 squares up from O*x*.

We always use the same method to describe the position of a point:
start from O, *first* move *along* and *then* up.

We can now shorten the description of the position of the point A to the number pair (3, 5).

228

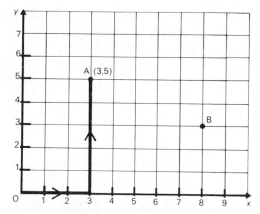

The number pair (3, 5) is referred to as the coordinates of A.
The first number, 3, is called the *x*-coordinate of A.
The second number, 5, is called the *y*-coordinate of A.

Now consider another point B

whose *x*-coordinate is 8
and whose *y*-coordinate is 3.

If we simply refer to   the point B(8, 3)

this tells us all that we need to know about the position of B.

The origin is the point (0, 0).

EXERCISE 16a   **1.**   Write down the coordinates of the points A, B, C, D, E, F, G and H.

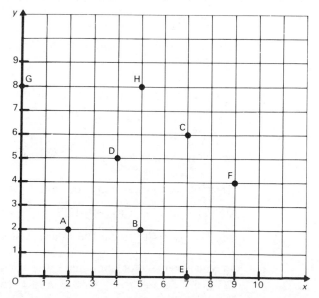

2. Draw a set of axes of your own. Give them scales from 0 to 10. Mark the following points and label each point with its own letter:

A(2, 8) B(4, 9) C(7, 9) D(8, 7) E(8, 6) F(9, 4) G(8, 4)
H(7, 3) I(5, 3) J(7, 2) K(7, 1) L(4, 2) M(2, 0) N(0, 2)

Now join your points together in alphabetical order and join A to N.

3. Draw a set of axes and give them scales from 0 to 10. Mark the following points:

A(2, 5) B(7, 5) C(7, 4) D(8, 4) E(8, 3) F(9, 3) G(9, 2)
H(6, 2) I(6, 1) J(7, 1) K(7, 0) L(5, 0) M(5, 2) N(4, 2)
P(4, 0) Q(2, 0) R(2, 1) S(3, 1) T(3, 2) U(0, 2) V(0, 3)
W(1, 3) X(1, 4) Y(2, 4)

Now join your points together in alphabetical order and join A to Y.

4. Mark the following points on your own set of axes:

A(2, 7) B(8, 7) C(8, 1) D(2, 1)

Join A to B, B to C, C to D and D to A. What is the name of the figure ABCD?

5. Mark the following points on your own set of axes:

A(2, 2) B(8, 2) C(5, 5)

Join A to B, B to C and C to A. What is the name of the figure ABC?

6. Mark the following points on your own set of axes:

A(4, 0) B(6, 0) C(6, 4) D(4, 4)

Join A to B, B to C, C to D and D to A. What is the name of the figure ABCD?

7. Mark the following points on your own set of axes:

A(5, 2) B(8, 5) C(5, 8) D(2, 5)

Join the points to make the figure ABCD. What is ABCD?

8. On your own set of axes mark the points A(8, 4), B(8, 8) and C(14, 6). Join A to B, B to C and C to A.
Describe the figure ABC.

9. Draw a simple pattern of your own on squared paper but do not show it to anyone. Write down the coordinates of each point and give this set of coordinates to your partner. See if your partner can now draw your diagram.

Questions 10 to 15 refer to the points A(1, 7), B(5, 0) and C(0, 14):

10. Write down the *x*-coordinate of the point B.

11. Write down the *y*-coordinate of the point A.

12. Write down the *x*-coordinate of the point C.

13. Write down the *x*-coordinate of the point A.

14. Write down the *y*-coordinate of the point C.

15. Write down the *y*-coordinate of the point B.

Questions 16 to 21 refer to points in the following diagram:

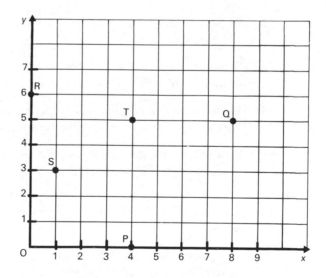

16. Write down the *y*-coordinate of the point T.

17. Write down the *x*-coordinate of the point P.

18. Write down the *x*-coordinate of the point S.

19. Write down the *y*-coordinate of the point R.

20. Write down the *y*-coordinate of the point Q.

21. Write down the *x*-coordinate of the point R.

Questions 22 to 25 refer to the following diagrams:

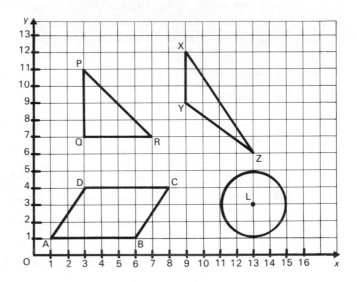

22. Write down the coordinates of the vertices X, Y and Z of triangle XYZ.

23. Write down the coordinates of the vertices of the isosceles triangle PQR. Write down the lengths of the two equal sides.

24. Write down the coordinates of the vertices of the parallelogram ABCD. How long is AB? How long is DC?

25. Write down the coordinates of the centre, L, of the circle. What is the diameter of this circle?

For each of the following questions you will need to draw your own set of axes:

26. The points A(2, 1), B(6, 1) and C(6, 5) are three corners of a square ABCD. Mark the points A, B and C. Find the point D and write down the coordinates of D.

27. The points A(2, 1), B(2, 3) and C(7, 3) are three vertices of a rectangle ABCD. Mark the points and find the point D. Write down the coordinates of D.

28. The points A(1, 4), B(4, 7) and C(7, 4) are three vertices of a square ABCD. Mark the points A, B and C and find D. Write down the coordinates of D.

29. Mark the points A(2, 4) and B(8, 4). Join A to B and find the point C which is the midpoint (the exact middle) of the line AB. Write down the coordinates of C.

30. Mark the points P(3, 5) and Q(3, 9). Join P and Q and mark the point R which is the midpoint of PQ. Write down the coordinates of R.

31. Mark the points A(0, 5) and B(4, 1). Find the coordinates of the midpoint of AB.

QUADRILATERALS

A quadrilateral has four sides. No two of the sides need be equal and no two of the sides need be parallel.

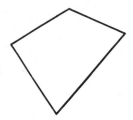

There are, however, some special quadrilaterals, such as a square, which have some sides parallel and/or some sides equal.

EXERCISE 16b If you are not sure whether two lines are equal, *measure them*.

If you are not sure whether two lines are parallel, *measure the corresponding angles*.

1. The Square

A(3, 2), B(11, 2), C(11, 10) and D(3, 10) are the four corners of a square. Mark these points on your own set of axes and then draw the square ABCD.

a) Write down, as a number of sides of squares, the lengths of the sides AB, BC, CD and DA.

b) Which side is parallel to AB? Are BC and AD parallel?

c) What is the size of each angle of the square?

2. The Rectangle

A(2, 2), B(2, 7), C(14, 7) and D(14, 2) are the vertices of a rectangle ABCD. Draw the rectangle ABCD on your own set of axes.

a) Write down the sides which are equal in length.

b) Write down the pairs of sides which are parallel.

c) What is the size of each angle of the rectangle?

3. The Rhombus

A(8, 1), B(11, 7), C(8, 13) and D(5, 7) are the vertices of a rhombus ABCD. Draw the rhombus on your own set of axes.

a) Write down the sides which are equal in length.
b) Write down the pairs of sides which are parallel.
c) Measure the angles of the rhombus. Are any of the angles equal?

4. The Parallelogram

A(2, 2), B(14, 2), C(17, 7) and D(5, 7) are the vertices of a parallelogram. Draw the parallelogram on your own set of axes.

a) Write down which sides are equal in length.
b) Write down which sides are parallel.
c) Measure the angles of the parallelogram. Write down which, if any, of the angles are equal.

5. The Trapezium

A(1, 1), B(12, 1), C(10, 5) and D(5, 5) are the vertices of a trapezium. Draw the trapezium on your own set of axes.

a) Write down which, if any, of the sides are the same length.
b) Write down which, if any, of the sides are parallel.
c) Write down which, if any, of the angles are equal.

PROPERTIES OF THE SIDES AND ANGLES OF THE SPECIAL QUADRILATERALS

We can summarize our investigations in the last exercise as follows:

In a square	all four sides are the same length
	both pairs of opposite sides are parallel
	all four angles are right angles.
In a rectangle	both pairs of opposite sides are the same length
	both pairs of opposite sides are parallel
	all four angles are right angles
In a rhombus	all four sides are the same length
	both pairs of opposite sides are parallel
	the opposite angles are equal.
In a parallelogram	the opposite sides are the same length
	the opposite sides are parallel
	the opposite angles are equal.
In a trapezium	just one pair of opposite sides are parallel.

EXERCISE 16c In the following questions the points A, B, C and D are the vertices of a quadrilateral. Draw the figure ABCD on your own set of axes and write down which type of quadrilateral it is.

1. A(2, 4)   B(7, 4)   C(8, 7)   D(3, 7)

2. A(2, 2)   B(6, 0)   C(7, 2)   D(3, 4)

3. A(2, 2)   B(7, 2)   C(5, 5)   D(3, 5)

4. A(2, 0)   B(6, 0)   C(6, 4)   D(2, 4)

5. A(1, 1)   B(4, 0)   C(4, 6)   D(1, 3)

6. A(3, 1)   B(6, 3)   C(3, 5)   D(0, 3)

7. A(1, 3)   B(4, 1)   C(6, 4)   D(3, 6)

8. A(2, 4)   B(3, 7)   C(9, 5)   D(8, 2)

9. A(3, 1)   B(5, 1)   C(3, 5)   D(1, 5)

10. A(0, 0)   B(5, 0)   C(8, 4)   D(3, 4)

NEGATIVE COORDINATES

If A(2, 0), B(4, 2) and C(6, 0) are three corners of a square ABCD, we can see that the fourth corner, D, is two squares below the *x*-axis.

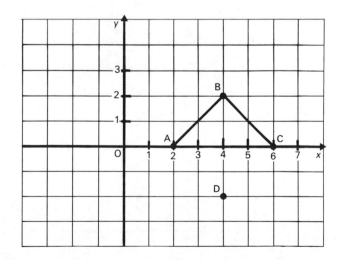

To describe the position of D we need to extend the scale on the *y*-axis below zero. To do this we use the numbers −1, −2, −3, −4, ... These are called negative numbers.

In the same way we can use the negative numbers −1, −2, −3, ... to extend the scale on the *x*-axis to the left of zero.

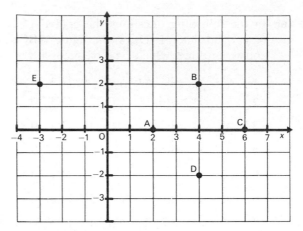

The *y*-coordinate of the point D is written −2 and is called "negative 2".

The *x*-coordinate of the point E is written −3 and is called "negative 3".

The numbers 1, 2, 3, 4 ... are called positive numbers. They could be written as +1, +2, +3, +4, ... but we do not usually put the + sign in.

Now D is   4 squares to the right of O so its *x*-coordinate is 4

    and   2 squares below the *x*-axis so its *y*-coordinate is −2,

                   D is the point (4, −2)

    E is   3 squares to the left of O so its *x*-coordinate is −3

    and   2 squares up from O so its *y*-coordinate is 2,

                   E is the point (−3, 2)

EXERCISE 16d

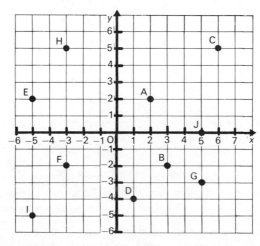

Use this diagram for questions 1 and 2.

1. Write down the *x*-coordinate of each of the points A, B, C, D, E, F, G, H, I, J and O (the origin).

2. Write down the *y*-coordinate of each of the points A, B, C, D, E, H, I and J.

The point Q has a *y*-coordinate of − 10. How many squares above or below the *x*-axis is the point Q?

Q is 10 squares below the *x*-axis.

How many squares above or below the *x*-axis is each of the following points?

3. P: the *y*-coordinate is −5   6. B: the *y*-coordinate is 10

4. L: the *y*-coordinate is +3   7. A: the *y*-coordinate is 0

5. M: the *y*-coordinate is −1   8. D: the *y*-coordinate is −4

How many squares to the left or to the right of the *y*-axis is each of the following points?

9. Q: the *x*-coordinate is 3   12. S: the *x*-coordinate is −7

10. R: the *x*-coordinate is −5   13. V: the *x*-coordinate is 0

11. T: the *x*-coordinate is +2   14. G: the *x*-coordinate is −9

Write down the coordinates of the point A.

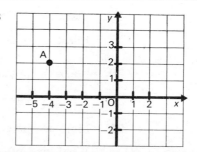

A is the point (−4, 2).

15. Write down the coordinates of the points A, B, C, D, E, F, G, H, I and J.

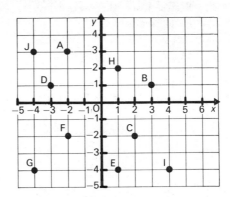

In questions 16 to 21 draw your own set of axes and scale each one from −5 to 5:

16. Mark the points   A(−3, 4) B(−1, 4) C(1, 3) D(1, 2) E(−1, 1) F(1, 0) G(1, −1) H(−1, −2) I(−3, −2).

Join the points in alphabetical order and join I to A.

17. Mark the points   A(4, −1) B(4, 2) C(3, 3) D(2, 3) E(2, 4) F(1, 4) G(1, 3) H(−2, 3) I(−3, 2) J(−3, −1).

Join the points in alphabetical order and join J to A.

18. Mark the points   A(2, 1) B(−1, 3) C(−3, 0) D(0, −2).

Join the points to make the figure ABCD. What is the name of the figure?

19. Mark the points   A(1, 3) B(−1, −1) C(3, −1).

Join the points to make the figure ABC and describe ABC.

20. Mark the points   A(−2, −1) B(5, −1) C(5, 2) D(−2, 2).

Join the points to make the figure ABCD and describe ABCD.

21. Mark the points   A(−3, 0) B(1, 3) C(0, −4).

What kind of triangle is ABC?

EXERCISE 16e Draw your own set of axes for each question in this exercise. Mark a scale on each axis from −10 to +10.

In questions 1 to 10 mark the points A and B and then find the length of the line AB:

1. A(2, 2) B(−4, 2)   **6.** A(5, −1) B(5, 6)

2. A(−2, −1) B(6, −1)   **7.** A(−2, 4) B(−7, 4)

3. A(−4, −4) B(−4, 2)   **8.** A(−1, −2) B(−8, −2)

4. A(1, −6) B(1, −8)   **9.** A(−3, 5) B(−3, −6)

5. A(3, 2) B(5, 2)   **10.** A(−2, −4) B(−2, 7)

In questions 11 to 20, the points A, B and C are three corners of a square ABCD. Mark the points and find the point D. Give the coordinates of D:

11. A(1, 1) B(1, −1) C(−1, −1)

12. A(1, 3) B(6, 3) C(6, −2)

13. A(3, 3) B(3, −1) C(−1, −1)

14. A(−2, −1) B(−2, 3) C(−6, 3)

15. A(−5, −3) B(−1, −3) C(−1, 1)

16. A(−3, −1) B(−3, 2) C(0, 2)

17. A(0, 4) B(−2, 1) C(1, −1)

18. A(1, 0) B(3, 2) C(1, 4)

19. A(−2, −1) B(2, −2) C(3, 2)

20. A(−3, −2) B(−5, 2) C(−1, 4)

In questions 21 to 30, mark the points A and B and the point C, the midpoint of the line AB. Give the coordinates of C:

21. A(2, 2) B(6, 2)   **26.** A(2, 1) B(6, 2)

22. A(2, 3) B(2, −5)   **27.** A(2, 1) B(−4, 5)

23. A(−1, 3) B(−6, 3)   **28.** A(−7, −3) B(5, 3)

24. A(−3, 5) B(−3, −7)   **29.** A(−3, 3) B(3, −3)

25. A(−1, −2) B(−9, −2)   **30.** A(−7, −3) B(5, 3)

STRAIGHT LINES

EXERCISE 16f **1.**

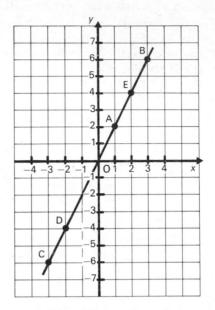

The points A, B, C, D and E are all on the same straight line.

a) Write down the coordinates of the points A, B, C, D and E.

b) F is another point on the same line. The x-coordinate of F is 5. Write down the y-coordinate of F.

c) G, H, I, J, K, L and M are also points on this line. Fill in the missing coordinates:

G(8, □)   H(10, □)   I(−4, □)   J(□, 12)   K(□, 18)
L(□, −10)   M(a, □).

2.

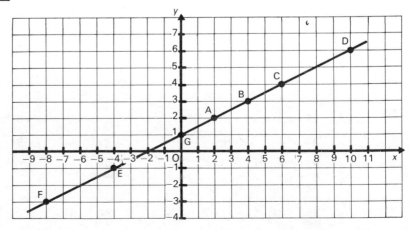

The points A, B, C, D, E, F and G are all on the same straight line.

a) Write down the coordinates of the points A, B, C, D, E, F and G.

b) How is the *y*-coordinate of each point related to its *x*-coordinate?

c) H is another point on this line. Its *x*-coordinate is 8; what is its *y*-coordinate?

d) I, J, K, L, M, N are further points on this line. Fill in the missing coordinates:

I(12, ☐) J(20, ☐) K(30, ☐) L(−12, ☐) M(☐, 9) N(*a*, ☐).

3.

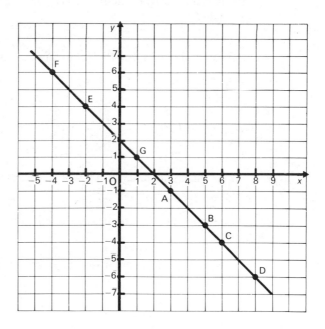

The points A, B, C, D, E, F and G are all on the same straight line.

a) Write down the coordinates of the points A, B, C, D, E, F and G.

b) H, I, J, K, L, M, N, P and Q are further points on the same line. Fill in the missing coordinates:

H(7, ☐)   I(10, ☐)   J(12, ☐)   K(20, ☐)   L(−7, ☐)
M(−9, ☐)   N(☐, 10)   P(☐, −8)   Q(☐, 12).

EXERCISE 16g In the following questions we are going to investigate the properties of the diagonals of the special quadrilaterals. You will need your own set of axes for each question. Mark a scale on each axis from −5 to +5. Mark the points A, B, C and D and join them to form the quadrilateral ABCD.

1. A(5, −2) B(2, 4) C(−3, 4) D(0, −2)
    a) What type of quadrilateral is ABCD?
    b) Join A to C and B to D. These are the diagonals of the quadrilateral. Mark with an E the point where the diagonals cross.
    c) Measure the diagonals. Are they the same length?
    d) Is E the midpoint of either, or both, of the diagonals?
    e) Measure the four angles at E. Do the diagonals cross at right angles?

Now repeat question 1 for the following points:

2. A(2, −2) B(2, 4) C(−4, 4) D(−4, −2)

3. A(2, −2) B(5, 4) C(−3, 4) D(−1, −2)

4. A(2, 0) B(0, 4) C(−2, 0) D(0, −4)

5. A(1, −4) B(1, −1) C(−5, −1) D(−5, −4)

6. Name the quadrilaterals in which the two diagonals are of equal length.

7. Name the quadrilaterals in which the diagonals cut at right angles.

8. Name the quadrilaterals in which the diagonals cut each other in half.

17 DIRECTED NUMBERS

USE OF POSITIVE AND NEGATIVE NUMBERS

Positive and negative numbers are collectively known as directed numbers.

Directed numbers can be used to describe any quantity that can be measured above or below a natural zero. For example, a distance of 50 m above sea level and a distance of 50 m below sea level could be written as $+50$ m and -50 m.

They can also be used to describe time before and after a particular event. For example, 5 seconds before the start of a race and 5 seconds after the start of a race could be written as -5 s and $+5$ s.

Directed numbers can also be used to describe quantities that involve one of two possible directions. For example, if a car is travelling north at 70 km/h and another car is travelling south at 70 km/h they can be described as going at $+70$ km/h and -70 km/h.

A familiar use of negative numbers is to describe temperatures. The freezing point of water is $0°$ centigrade (or Celsius) and a temperature of $5°$C below freezing point is written $-5°$C.

Most people would call $-5°$C "minus $5°$C" but we will call it "negative $5°$C" and there are good reasons for doing so because in mathematics "minus" means "take away".

A temperature of $5°$C above freezing point is called "positive $5°$C" and can be written as $+5°$C. Most people would just call it $5°$C and write it without the positive symbol.

> A number without any symbol in front of it is a positive number,
> i.e. 2 means $+2$
> and $+3$ can be written as 3

EXERCISE 17a Draw a centigrade thermometer and mark a scale on it from $-10°$ to $+10°$. Use your drawing to write the following temperatures as positive or negative numbers:

1. $10°$ above freezing point

2. $7°$ below freezing point

3. $3°$ below zero

4. $5°$ above zero

5. $8°$ below zero

6. freezing point

Write down, in words, the meaning of the following temperatures:

7. $-2\,°C$ **10.** $-10\,°C$

8. $+3\,°C$ **11.** $+8\,°C$

9. $4\,°C$ **12.** $0\,°C$

Which temperature is higher?

13. $+8°$ or $+10°$ **18.** $-2°$ or $-5°$

14. $12°$ or $3°$ **19.** $1°$ or $-1°$

15. $-2°$ or $+4°$ **20.** $+3°$ or $-5°$

16. $-3°$ or $-5°$ **21.** $-7°$ or $-10°$

17. $-8°$ or $2°$ **22.** $-2°$ or $-9°$

23. The contour lines on the map below show distances above sea level as positive numbers and distances below sea level as negative numbers.

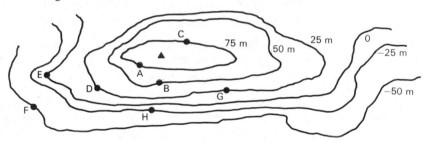

Write down in words the position relative to sea level of the points A, B, C, D, E, F, G and H.

In questions 24 to 34 use positive or negative numbers to describe the quantities.

A ball thrown up a distance of 5 m.

$+5\,m$

24. 5 seconds before blastoff of a rocket.

25. 5 seconds after blastoff of a rocket.

26. 50 p in your purse.

27. 50 p owed.

28. 1 minute before the train leaves the station.

29. A win of £50 on premium bonds.

30. A debt of £5.

31. Walking forwards five paces.

32. Walking backwards five paces.

33. The top of a hill which is 200 m above sea level.

34. A ball thrown down a distance of 5 m.

35. At midnight the temperature was −2 °C. One hour later it was 1° colder. What was the temperature then?

36. At midday the temperature was 18 °C. Two hours later it was 3° warmer. What was the temperature then?

37. A rockclimber started at +200 m and came a distance of 50 m down the rock face. How far above sea level was he then?

38. At midnight the temperature was −5 °C. One hour later it was 2° warmer. What was the temperature then?

39. At the end of the week my financial state could be described as −25 p. I was later given 50 p. How could I then describe my financial state?

40. Positive numbers are used to describe a number of paces forwards and negative numbers are used to describe a number of paces backwards. Describe where you are in relation to your starting point if you walk +10 paces followed by −4 paces.

EXTENDING THE NUMBER LINE

If a number line is extended beyond zero, negative numbers can be used to describe points to the left of zero and positive numbers are used to describe points to the right of zero.

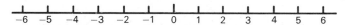

On this number line,   5 is to the *right* of 3

and we say that   5 is *greater* than 3

        or   5 > 3

Also   −2 is to the *right* of −4

and we say that   −2 is *greater* than −4

        or   −2 > −4

So "greater" means "higher up the scale".
(A temperature of $-2\,°C$ is higher than a temperature of $-4\,°C$.)

Now	2 is to the *left* of 6
and we say that	2 is *less* than 6
	or $2 < 6$
Also	-3 is to the *left* of -1
and we say that	-3 is *less* than -1
	or $-3 < -1$

So "less than" means "lower down the scale".

EXERCISE 17b   Draw a number line.
In questions 1 to 12 write either $>$ or $<$ between the two numbers:

1. 3   2   **5.** 1 -2 **9.** -3 -9

2. 5   1   **6.** -4 1   **10.** -7 3

3. -1 -4 **7.** 3 -2 **11.** -1 0

4. -3 -1 **8.** 5 -10 **12.** 1 -1

In questions 13 to 24 write down the next two numbers in the sequence:

13. 4, 6, 8   **17.** 9, 6, 3   **21.** 36, 6, 1

14. $-4, -6, -8$ **18.** $-4, -1, 2$ **22.** $-10, -8, -6$

15. 4, 2, 0   **19.** 5, 1, -3 **23.** $-1, -2, -4$

16. $-4, -2, 0$ **20.** 2, 4, 8   **24.** 1, 0, -1

ADDITION AND SUBTRACTION OF POSITIVE NUMBERS

If you were asked to work out $5-7$ you would probably say that it cannot be done. But if you were asked to work out where you would be if you walked 5 steps forwards and then 7 steps backwards, you would say that you were two steps behind your starting point.

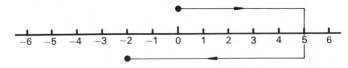

On the number line, $5-7$ means

          start at 0 and go 5 places to the right

          and then go 7 places to the left

So $5 - 7 = -2$

i.e. "minus" a positive number means   move to the left

and "plus" a positive number means   move to the right.

In this way $3+2-8+1$ can be shown on the number line as follows:

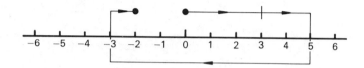

Therefore $3+2-8+1 = -2$.

EXERCISE 17c Find, using a number line if it helps:

1.	$3-6$	**6.**	$5+2$
2.	$5-2$	**7.**	$-2+3$
3.	$4-6$	**8.**	$-3+5$
4.	$5-7$	**9.**	$-5-7$
5.	$4-2$	**10.**	$-3+2$

$(+4)+(+3)$

           $(+4)+(+3) = 4+3$

                     $= 7$

$(+4)-(+3)$

           $(+4)-(+3) = 4-3$

                     $= 1$

11.	$(+3)+(+2)$	**14.**	$-(+3)+(+2)$
12.	$(+2)-(+4)$	**15.**	$-(+1)-(+5)$
13.	$(+5)-(+7)$	**16.**	$5-2+3$

17. $7-9+4$

18. $5-11+3$

19. $10-4-9$

20. $3+6-10$

21. $-4+2+5$

22. $-3+1-4$

23. $5-6-9$

24. $-3-4+2$

25. $-2-3+9$

26. $(+3)+(+4)-(+1)$

27. $(+2)-(+5)+(+6)$

28. $(+9)-(+7)-(+2)$

29. $-(+3)+(+5)-(+5)$

30. $-(+8)-(+4)+(+7)$

What number does x represent if $x+2 = 1$?

$$-1+2 \quad \text{is} \quad 1$$

$$\text{so} \quad x = -1$$

Find the value of x:

31. $x+4 = 5$

32. $x-2 = 0$

33. $x+2 = 4$

34. $x+2 = 0$

35. $2+x = 1$

36. $3+x = 1$

37. $5-x = 4$

38. $5+x = 7$

39. $9-x = 4$

40. $x-6 = 10$

ADDITION AND SUBTRACTION OF NEGATIVE NUMBERS

Most of you will have some money of your own, from pocket money and other sources. Many of you will have borrowed money at some time.

At any one time you have a *balance* of money, i.e. the total sum that you own or owe!

If you own £2 and you borrow £4, your balance is a debt of £2.
We can write this as

$$(+2) + (-4) = (-2)$$

or as

$$2 + (-4) = -2$$

But

$$2-4 = -2$$

∴

$$+(-4) \quad \text{means} \quad -4$$

If you owe £2 and then take away that debt, your balance is zero. We can write this as

$$(-2) - (-2) = 0$$

You can pay off a debt on your balance only if someone gives you £2. So subtracting a negative number is equivalent to adding a positive number, i.e. $-(-2)$ is equivalent to $+2$.

$$-(-2) \quad \text{means} \quad +2$$

EXERCISE 17d Find:

> $2+(-1)$
>
> $$2+(-1) = 2-1$$
> $$= 1$$

> $-3-(-4)$
>
> $$-3-(-4) = -3+4$$
> $$= 1$$

1. $3+(-1)$	**6.** $-2-(-5)$	**11.** $-7+(-7)$
2. $5+(-8)$	**7.** $4+(-7)$	**12.** $-3-(-3)$
3. $4-(-3)$	**8.** $-3-(-9)$	**13.** $+4+(-4)$
4. $-1-(-4)$	**9.** $-4+(-10)$	**14.** $+2-(-4)$
5. $-2+(-7)$	**10.** $2-(-8)$	**15.** $-3+(-3)$

> $2+(-1)-(-4)$
>
> $$2+(-1)-(-4) = 2-1+4$$
> $$= 5$$

16. $5+(-1)-(-3)$	**21.** $9+(-5)-(-9)$
17. $(-1)+(-1)+(-1)$	**22.** $8-(-7)+(-2)$
18. $4-(-2)+(-4)$	**23.** $10+(-9)+(-7)$
19. $-2-(-2)+(-4)$	**24.** $12+(-8)-(-4)$
20. $6-(-7)+(-8)$	**25.** $9+(-12)-(-4)$

ADDITION AND SUBTRACTION OF DIRECTED NUMBERS ———

We can now use the following rules:

$$+(+a) = +a \qquad \text{and} \qquad -(+a) = -a$$
$$+(-a) = -a \qquad \text{and} \qquad -(-a) = +a$$

EXERCISE 17e Find:

1. $3+(-2)$

2. $-3-(+2)$

3. $6-(-3)$

4. $4+(+4)$

5. $-5+(-7)$

6. $9-(+2)$

7. $7+(-3)$

8. $8+(+2)$

9. $10-(-5)$

10. $-2-(-4)$

11. $12+(-7)$

12. $-4-(+8)$

13. $3-(-2)$

14. $-5+(-4)$

15. $8+(-7)$

16. $4-(-5)$

17. $7+(-3)-(+5)$

18. $2-(-4)+(-6)$

19. $5+(-2)-(+1)$

20. $8-(-3)+(+5)$

21. $7+(-4)-(-2)$

22. $3-(+2)+(-5)$

23. $-9+(-2)-(-3)$

24. $8+(+9)-(-2)$

25. $7+(-9)-(+2)$

26. $4+(-1)-(+7)$

27. $-3+(+5)-(-2)$

28. $-4+(+8)+(-7)$

29. $-9-(+4)-(-10)$

30. $-2-(+8)+(-9)$

$-8-(4-7)$

$$-8-(4-7) = -8-(-3) \qquad \text{(brackets first)}$$
$$= -8+3$$
$$= -5$$

---

31. $3-(4-3)$

32. $5+(7-9)$

33. $4+(8-12)$

34. $-3-(7-10)$

35. $6+(8-15)$

36. $(3-5)+2$

37. $5-(6-10)$ **41.** $(3-8)-(9-4)$

38. $(4-9)-2$ **42.** $(3-1)+(5-10)$

39. $(7+4)-15$ **43.** $(7-12)-(6-9)$

40. $8+(3-8)$ **44.** $(4-8)-(10-15)$

45. Add $(+7)$ to (-5).

46. Subtract 7 from -5.

47. Subtract (-2) from 1.

48. Find the value of "8 take away -10".

49. Add -5 to $+3$.

50. Find the sum of -3 and $+4$.

51. Find the sum of -8 and $+10$.

52. Subtract positive 8 from negative 7.

53. Find the sum of -3 and -3 and -3.

54. Find the value of twice negative 3.

55. Find the value of four times -2.

MULTIPLICATION OF DIRECTED NUMBERS

Consider 3×2.

This means $2 + 2 + 2 = 6$

So $4 \times (-3)$ means $(-3) + (-3) + (-3) + (-3) = -12$

Order does not matter when we multiply   (think of $2 \times 5 = 5 \times 2$).

Therefore we can say

$$(-3) \times 4 = 4 \times (-3) = -12$$

EXERCISE 17f

Find the following products:

a) $5 \times (-3)$ b) $(-3) \times 5$ c) $(+5) \times (-3)$ d) $(-3) \times (+5)$

a) $5 \times (-3) = -15$ b) $(-3) \times 5 = -15$

c) $(+5) \times (-3) = -15$ d) $(-3) \times (+5) = -15$

Find the following products:

1.	$6 \times (-4)$	**6.**	$(+4) \times (-3)$	**11.**	$8 \times (-2)$
2.	$7 \times (-2)$	**7.**	$(-6) \times (+8)$	**12.**	$(-9) \times 4$
3.	$(-8) \times 3$	**8.**	$(+1) \times (-5)$	**13.**	$(+6) \times (-7)$
4.	$(-6) \times 2$	**9.**	$6 \times (-1)$	**14.**	$(-5) \times 1$
5.	$9 \times (-3)$	**10.**	$(-1) \times (+5)$	**15.**	$5 \times (-2.5)$

DIVISION OF NEGATIVE NUMBERS BY POSITIVE NUMBERS

Since $2 \times 3 = 6$, $6 \div 3 = 2$.

In the same way, $(-3) \times 4 = -12$, so $(-12) \div 4 = -3$.

Notice that order *does* matter in division, e.g.

$$(-12) \div 4 = -3$$

but we shall see that $4 \div (-12) = \dfrac{-1}{3}$

EXERCISE 17g

Find   a) $(-9) \div 3$ b) $(-14) \div (+2)$ c) $\dfrac{-10}{2}$

a) $(-9) \div 3 = -3$

b) $(-14) \div (+2) = -7$

c) $\dfrac{-10}{2} = -5$

Find:

1.	$(-6) \div 2$	**5.**	$(-12) \div (+3)$	**9.**	$(-20) \div 4$
2.	$(-10) \div 5$	**6.**	$(-18) \div (+9)$	**10.**	$(-28) \div (+7)$
3.	$(-15) \div 3$	**7.**	$(-30) \div (+3)$	**11.**	$(-3) \div 3$
4.	$(-24) \div 6$	**8.**	$(-36) \div 12$	**12.**	$(-10) \div (+5)$

13.	$\dfrac{-8}{4}$	**15.**	$\dfrac{-16}{4}$	**17.**	$\dfrac{-36}{9}$
14.	$\dfrac{-12}{6}$	**16.**	$\dfrac{-27}{3}$	**18.**	$\dfrac{-30}{15}$

MIXED EXERCISES

EXERCISE 17h **1.** Which is the higher temperature, $-5°$ or $-8°$?

2. Write $<$ or $>$ between   a) -3 2   b) -2 -4.

Find:

3. $-4+6$ **7.** $-2+(-3)-(-5)$

4. $3+2-10$ **8.** $4-(2-3)$

5. $2+(-4)$ **9.** $6\times(-4)$

6. $3-(-1)$ **10.** $-36\div3$

EXERCISE 17i **1.** Which is the lower temperature, $0°$ or $-3°$?

2. Write $<$ or $>$ between   a) 3 -4 b) -7 -10.

Find:

3. $2-8$ **7.** $3+(5-8)$

4. $3-9+4$ **8.** $-2-(4-9)$

5. $(+2)-(-3)$ **9.** $(-3)\times4$

6. $(-4)-(-5)$ **10.** $(-12)\div6$

18 INTRODUCING ALGEBRA

THE IDEA OF EQUATIONS

"I think of a number, and take away 3; the result is 7."

We can see the number must be 10.

Using a letter to stand for the unknown number we can write the first sentence as an equation:

$$x - 3 = 7$$

Then if $x = 10$ 　　　　　 $10 - 3 = 7$

so $x = 10$ fits the equation.

EXERCISE 18a Form equations to illustrate the following statements and find the unknown numbers:

> I think of a number, add 4 and the result is 10.
>
> The equation is $x + 4 = 10$.
> The number is 6

1. I think of a number, subtract 3 and get 4.

2. I think of a number, add 1 and the result is 3.

3. If a number is added to 3 we get 9.

4. If 5 is subtracted from a number we get 2.

> I think of a number, multiply it by 3 and the result is 12.
>
> The equation is $3x = 12$ 　　　($3x$ means $3 \times x$).
> The number is 4.

5. I think of a number, double it and get 8.

6. If a number is multiplied by 7 the result is 14.

7. When we multiply a number by 3 we get 15.

8. 6 times an unknown number gives 24.

Write sentences to show the meaning of the following equations:

$4x = 20$

$4x = 20$ means   4 times an unknown number gives 20, or, I think of a number, multiply it by 4 and the result is 20.

9. $3x = 18$ **13.** $5+x = 7$

10. $x+6 = 7$ **14.** $x-4 = 1$

11. $x-2 = 9$ **15.** $4x = 8$

12. $5x = 20$ **16.** $x+1 = 4$

SOLVING EQUATIONS

Some equations need an organised approach, not guesswork.

Imagine a balance:

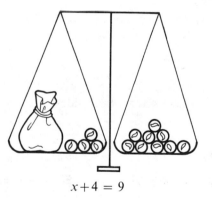

on this side there is a bag containing an unknown number of marbles, say x marbles, and 4 loose marbles

on this side, there are 9 separate marbles, balancing the marbles on the other side

$$x+4 = 9$$

Take 4 loose marbles from each side, so that the two sides still balance.

$$x = 5$$

We write: $x + 4 = 9$

Take 4 from both sides $x = 5$

When we have found the value of x we have *solved the equation*.

As a second example suppose that:

on this side there is a
bag that originally held
x marbles but now has
2 missing

on this side, there are
5 loose marbles

$$x - 2 = 5$$

We can make the bag complete by putting back 2 marbles but, to keep the balance, we must add 2 marbles to the right-hand side also.

So we write $x - 2 = 5$

Add 2 to both sides $x = 7$

> Whatever you do to one side of an equation you must also do to the other side.

EXERCISE 18b Solve the following equations:

> $y + 4 = 6$
>
> $\qquad\qquad y + 4 = 6$
>
> Take 4 from both sides $y = 2$

1. $x + 7 = 15$ **5.** $a + 3 = 7$ **9.** $a + 1 = 6$

2. $x + 9 = 18$ **6.** $x + 4 = 9$ **10.** $a + 8 = 15$

3. $10 + y = 12$ **7.** $a + 5 = 11$ **11.** $7 + c = 10$

4. $2 + c = 9$ **8.** $9 + a = 15$ **12.** $c + 2 = 3$

Some equations may have negative answers.

> $x + 8 = 6$
>
> $\qquad\qquad x + 8 = 6$
>
> Take 8 from both sides $x = -2$

13. $x + 4 = 2$ **15.** $3 + a = 2$ **17.** $4 + w = 2$

14. $x + 6 = 1$ **16.** $s + 3 = 2$ **18.** $c + 6 = 2$

$$x-6 = 2$$

$$x-6 = 2$$

Add 6 to both sides $x = 8$

19. $x-6 = 4$ **23.** $c-8 = 1$ **27.** $a-4 = 8$

20. $a-2 = 1$ **24.** $x-5 = 7$ **28.** $x-3 = 0$

21. $y-3 = 5$ **25.** $s-4 = 1$ **29.** $c-1 = 1$

22. $x-4 = 6$ **26.** $x-9 = 3$ **30.** $y-7 = 2$

EXERCISE 18c Sometimes the letter term is on the right-hand side instead of the left. Solve the following equations:

$$3 = x-4$$

$$3 = x-4$$

Add 4 to both sides $7 = x$ ($7 = x$ is the same as $x = 7$)

$$x = 7$$

1. $4 = x+2$ **3.** $7 = a+4$ **5.** $1 = c-2$

2. $6 = x-3$ **4.** $6 = x--7$ **6.** $5 = s+2$

7. $x+3 = 10$ **11.** $6+c = 10$ **15.** $x+6 = 5$

8. $9+x = 4$ **12.** $d+4 = 1$ **16.** $x+3 = 15$

9. $c+4 = 4$ **13.** $7 = x+3$ **17.** $y-6 = 4$

10. $3 = b+2$ **14.** $x+1 = 9$ **18.** $x-7 = 4$

19. $6 = x-4$ **23.** $10 = a-1$ **27.** $x-3 = 6$

20. $x-4 = 2$ **24.** $c-7 = 9$ **28.** $c-7 = 10$

21. $x-9 = 2$ **25.** $x-4 = 8$ **29.** $4 = b-1$

22. $x-1 = 4$ **26.** $y-1 = 9$ **30.** $x-4 = 12$

31. $y-9 = 14$ **33.** $x+1 = 8$ **35.** $x+8 = 1$

32. $2 = z-2$ **34.** $x-1 = 8$ **36.** $x-8 = 1$

37. $c+5 = 9$ **39.** $1 = c-3$ **41.** $3+z = 1$

38. $d-3 = 1$ **40.** $1 = c+3$ **42.** $z+3 = 5$

MULTIPLES OF x

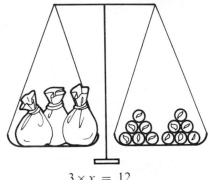

Imagine that on this side of the scales there are 3 bags each containing an equal unknown number of marbles, say x in each

On this side there are 12 loose marbles

$$3 \times x = 12$$

$$3x = 12$$

We can keep the balance if we divide the contents of each scale pan by 3.

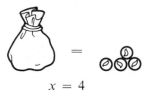

$$x = 4$$

EXERCISE 18d Solve the following equations:

$6x = 12$

$6x = 12$

Divide both sides by 6 $x = 2$

$3x = 7$

$3x = 7$

Divide both sides by 3 $x = \frac{7}{3}$

$x = 2\frac{1}{3}$

1. $5x = 10$	**5.** $4b = 16$	**9.** $5p = 7$
2. $3x = 9$	**6.** $4c = 9$	**10.** $2x = 40$
3. $2x = 5$	**7.** $3a = 1$	**11.** $7y = 14$
4. $7x = 21$	**8.** $6z = 18$	**12.** $6a = 3$

13. $6x = 36$	**17.** $5z = 9$	**21.** $4y = 3$
14. $6x = 6$	**18.** $2y = 7$	**22.** $5x = 6$
15. $6x = 1$	**19.** $3x = 27$	**23.** $2z = 10$
16. $5z = 10$	**20.** $8x = 16$	**24.** $7x = 1$

MIXED OPERATIONS

EXERCISE 18e Solve the following equations:

1. $x+4 = 8$	**5.** $5y = 6$	**9.** $2x = 11$
2. $x-4 = 8$	**6.** $4x = 12$	**10.** $x-2 = 11$
3. $4x = 8$	**7.** $4+x = 12$	**11.** $12 = x+4$
4. $5+y = 6$	**8.** $x-4 = 12$	**12.** $x-12 = 4$

13. $8 = c+2$	**17.** $7y = 2$	**21.** $3 = a-4$
14. $3x = 10$	**18.** $3+x = 2$	**22.** $x+3 = 5$
15. $20 = 4x$	**19.** $3x = 8$	**23.** $3x = 5$
16. $5+x = 4$	**20.** $x+6 = 1$	**24.** $z-5 = 6$

25. $c+5 = 5$	**27.** $a+5 = 25$	**29.** $a-25 = 5$
26. $5a = 25$	**28.** $a-5 = 25$	**30.** $25a = 5$

TWO OPERATIONS

EXERCISE 18f The aim is to get the letter term on its own.

Solve the following equations:

$7 = 3x - 5$

$7 = 3x - 5$

Add 5 to both sides $12 = 3x$

Divide both sides by 3 $4 = x$

i.e. $x = 4$

$2x + 3 = 5$

$2x + 3 = 5$

Take 3 from both sides $2x = 2$

Divide both sides by 2 $x = 1$

(It is possible to check whether your answer is correct. We can put $x = 1$ in the left-hand side of the equation and see if we get the same value on the right-hand side.)

Check: If $x = 1$, left-hand side $= 2 \times 1 + 3 = 5$

Right-hand side $= 5$, so $x = 1$ fits the equation.

1. $6x + 2 = 26$ **8.** $6 = 2x - 4$ **15.** $20 = 12x - 4$

2. $4x + 7 = 19$ **9.** $5z + 9 = 4$ **16.** $9x + 1 = 28$

3. $17 = 7x + 3$ **10.** $3x - 4 = 4$ **17.** $9 = 8x - 15$

4. $4x - 5 = 19$ **11.** $3x + 4 = 25$ **18.** $8 = 8 + 3z$

5. $7x + 1 = 22$ **12.** $2x + 15 = 25$ **19.** $6x + 7 = 1$

6. $3a + 12 = 12$ **13.** $13 = 3x + 4$ **20.** $5x - 4 = 5$

7. $10 = 10x - 50$ **14.** $5z - 9 = 16$ **21.** $15 = 1 + 7x$

22. $9x - 4 = 14$ **27.** $3a + 4 = 1$ **32.** $10x - 6 = 24$

23. $3x - 2 = 3$ **28.** $2x + 6 = 6$ **33.** $5x - 7 = 4$

24. $4 + 5x = -5$ **29.** $19x - 16 = 22$ **34.** $10 + 2x = -2$

25. $7 = 2z + 6$ **30.** $3x + 1 = 11$ **35.** $8 = 3x + 7$

26. $5 = 7x - 23$ **31.** $16 = 7x - 1$ **36.** $9 = 6a - 27$

37. $6x + 1 = -5$ **39.** $2x + 4 = 14$ **41.** $7 = 1 - 2x$

38. $4z + 3 = 4$ **40.** $3 = 7x - 3$ **42.** $8x + 11 = 3$

PROBLEMS

EXERCISE 18g Form equations and solve the problems:

> I think of a number, double it and add 3. The result is 15.
> What is the number?
>
> Let the number be x $2x + 3 = 15$
>
> Take 3 from both sides $2x = 12$
>
> Divide both sides by 2 $x = 6$
>
> The number is 6.

> The side of a square is x cm. Its perimeter is 20 cm.
> Find x.
>
>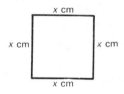
>
> The perimeter is $x + x + x + x$ cm, which is $4x$ cm
>
> $$4x = 20$$
>
> Divide both sides by 4 $x = 5$

1. I think of a number, multiply it by 4 and subtract 8. The result is 20. What is the number?

2. I think of a number, multiply it by 6 and subtract 12. The result is 30. What is the number?

3. I think of a number, multiply it by 3 and add 6. The result is 21. What is the number?

4. When 8 is added to an unknown number the result is 10. What is the number?

5. I think of a number, multiply it by 3 and add the result to 7. The total is 28. What is the number?

6. The sides of a rectangle are x cm and 3 cm. Its perimeter is 24 cm. Find x.

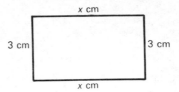

x cm

3 cm   3 cm

x cm

7. The lengths of the three sides of a triangle are x cm, x cm and 6 cm. Its perimeter is 20 cm. Find x.

8. Mary and Jean each have x sweets and Susan has 10 sweets. Amongst them they have 24 sweets. What is x?

9. Three boys had x sweets each. Amongst them they gave 9 sweets to a fourth boy and then found that they had 18 sweets left altogether. Find x.

10. I have two pieces of ribbon each x cm long and a third piece 9 cm long. Altogether there are 31 cm of ribbon. What is the length of each of the first two pieces?

EQUATIONS WITH LETTER TERMS ON BOTH SIDES

Some equations have letter terms on both sides. Consider the equation

$$5x + 1 = 2x + 9$$

We want to have a letter term on one side only so we need to take $2x$ from both sides. This gives

$$3x + 1 = 9$$

and we can go on to solve the equation as before.

Notice that we want the letter term on the side which has the greater number of x's to start with.

If we look at the equation

$$9 - 4x = 2x + 4$$

we can see that there is a lack of x's on the left-hand side, so there are more x's on the right-hand side. Add $4x$ to both sides and then the equation becomes

$$9 = 6x + 4$$

and we can go on as before.

EXERCISE 18h Deal with the letters first, then the numbers.

Solve the following equations:

$5x+2 = 2x+9$

$$5x+2 = 2x+9$$

Take $2x$ from both sides $\quad 3x+2 = 9$

Take 2 from both sides $\quad\quad 3x = 7$

Divide both sides by 3 $\quad\quad x = \dfrac{7}{3} = 2\dfrac{1}{3}$

1.	$3x+4 = 2x+8$	**5.**	$7x+3 = 3x+31$
2.	$x+7 = 4x+4$	**6.**	$6z+4 = 2z+1$
3.	$2x+5 = 5x-4$	**7.**	$7x-25 = 3x-1$
4.	$3x-1 = 5x-11$	**8.**	$11x-6 = 8x+9$

$9+x = 4-4x$

$$9+x = 4-4x$$

Add $4x$ to both sides $\quad 9+5x = 4$

Take 9 from both sides $\quad\quad 5x = -5$

Divide both sides by 5 $\quad\quad x = -1$

Check: If $x = -1$, left-hand side $= 9+(-1)$

$$= 8$$

right-hand side $= 4-(-4)$

$$= 8$$

So $\quad x = -1 \quad$ is the solution.

9.	$4x-3 = 39-2x$	**13.**	$5x-6 = 3-4x$
10.	$5+x = 17-5x$	**14.**	$12+2x = 24-4x$
11.	$7-2x = 4+x$	**15.**	$32-6x = 8+2x$
12.	$24-2x = 5x+3$	**16.**	$9-3x = -5+4x$

$9-3x = 15-4x$

$$9-3x = 15-4x$$

(notice that there is a greater lack of x's on the right)

Add $4x$ to both sides $9+x = 15$

Take 9 from both sides $x = 6$

17.	$5-3x = 1-x$	**21.**	$16-6x = 1-x$
18.	$16-2x = 19-5x$	**22.**	$4-3x = 1-4x$
19.	$6-x = 12-2x$	**23.**	$4-2x = 8-5x$
20.	$-2-4x = 6-2x$	**24.**	$3-x = 5-3x$

25.	$6-3x = 4x-1$	**29.**	$13-4x = 4x-3$
26.	$4z+1 = 6z-3$	**30.**	$7x+6 = x-6$
27.	$3-6x = 6x-3$	**31.**	$6-2x = 9-5x$
28.	$8-4x = 14-7x$	**32.**	$3-2x = 3+x$

$3-2x = 5$

$$3-2x = 5$$

(the left-hand side has a lack of x's)

Add $2x$ to both sides $3 = 5+2x$

Take 5 from both sides $-2 = 2x$

Divide both sides by 2 $x = -1$

33.	$13-4x = 5$	**35.**	$6 = 8-3x$
34.	$6 = 2-2x$	**36.**	$0 = 6-2x$

37.	$9x+4 = 3x+1$	**42.**	$5-3x = 2$
38.	$2x+3 = 12x$	**43.**	$6+3x = 7-x$
39.	$7-2x = 3-6x$	**44.**	$5-2x = 4x-7$
40.	$3x-6 = 6-x$	**45.**	$5x+3 = -7-x$
41.	$-4x-5 = -2x-10$	**46.**	$4-3x = 0$

SIMPLIFYING EXPRESSIONS

Like Terms

Consider $3x+5x-4x+2x$.
This is called an *expression* and can be simplified to $6x$.
$3x$, $5x$, $4x$ and $2x$ are all *terms* in this expression. Each term contains x. They are of the same type and are called *like terms*.

EXERCISE 18i Simplify:

$$4h-6h+7h-h$$
$$4h-6h+7h-h = 4h$$

1. $3x+x+4x+2x$

2. $3x-x+4x-2x$

3. $-6x+8x$

4. $6-1+4-7$

5. $-8x+6x$

6. $9y-3y+2y$

7. $2-3+9-1$

8. $-16-3-4$

9. $-3+5-1$

10. $-2x-x+3x$

Unlike Terms

$3x+2x-7$ can be simplified to $5x-7$, and $5x-2y+4x-3y$ can be simplified to $9x-5y$.

Terms containing x are different from terms without an x. They are called *unlike terms* and cannot be collected. Similarly $9x$ and $5y$ are unlike terms; therefore $9x-5y$ cannot be simplified.

EXERCISE 18j Simplify:

$$3x+4-7-2x+4x$$
$$3x+4-7-2x+4x = 5x-3$$

$$2x+4y-x-5y$$
$$2x+4y-x-5y = x-y$$

1. $2x+4+3+5x$

2. $2x-4+3x+9$

3. $5x-2-3-x$

4. $4a+5c-6a$

5. $6x-5y+2x+3y$

6. $6x+5y+2x+3y$

7. $6x+5y+2x-3y$

8. $6x+5y-2x+3y$

9. $4x+1+3x+2+x$

10. $6x-9+2x+1$

11. $7x-3-9-4x$

12. $9x+3y-10x$

13. $2x-6y-8x$

14. $7-x-6-3x$

15. $8-1-7x+2x$

16. $9x-1+4-11x$

17. $6x-5y+2x+3y+2x$

18. $6x-5y-2x-3y+7x-y$

19. $30x+2-15x-6+4$

20. $-2z+3x-4y+6z+x-3y$

21. $4x+3y-4+6x-2y-7-x$

22. $7x+3-9-9x+2x-6+11$

EQUATIONS CONTAINING LIKE TERMS

If there are a lot of terms in an equation, first collect the like terms on each side separately.

EXERCISE 18k Solve the following equations:

$$2x+3-x+5 = 3x+4x-6$$
$$2x+3-x+5 = 3x+4x-6$$
$$x+8 = 7x-6$$

Take x from both sides $\quad 8 = 6x-6$

Add 6 to both sides $\quad 14 = 6x$

Divide both sides by 6 $\quad \frac{14}{6} = x$

$$x = \frac{7}{3} = 2\frac{1}{3}$$

1. $3x+2+2x = 7$

2. $7+3x-6 = 4$

3. $6 = 5x+2-4x$

4. $9+4 = 3x+4x$

5. $3x+8-5x = 2$

6. $3x+2x-4x = 6$

7. $7 = 2-3+4x$

8. $5x+x-6x+2x = 9$

9. $5+x-4x+x = 1$

10. $6x = x+2-7-1$

11. $5x+6+3x = 10$

12. $8 = 7-11+6x$

13. $7+2x = 12x-7x+2$

14. $1-4-3+2x = 3x$

15. $3x-4x-x = x-6$

16. $2-4x-x = x+8$

17. $-2x+x = 3x-12$ **19.** $4-x-2-x = x$

18. $-4+x-2-x = x$ **20.** $4-x-2+x = x$

21. $3x+1+2x = 6$ **26.** $2x+7-4x+1 = 4$

22. $4x-2+6x-4 = 64$ **27.** $6-3x-5x-1 = 10$

23. $2x+7-x+3 = 6x$ **28.** $6x+3+6 = x-4-2$

24. $6-2x-4+5x = 17$ **29.** $x-3+7x+9 = 10$

25. $9x-6-x-2 = 0$ **30.** $15x+2x-6x-9x = 20$

MIXED EXERCISES

EXERCISE 18l **1.** Solve the equation $3x+2 = 4$.

2. I think of a number, add 4 and the result is 10. Form an equation and solve it to find the number I thought of.

3. Solve the equation $6x+2 = 3x+8$.

4. Solve the equation $4x-2 = -6$.

5. Simplify $4x-3y+5x+2y$.

6. Solve the equation $4x+2-x = 6$.

EXERCISE 18m **1.** Solve the equation $4x-5 = 3$.

2. Simplify $3c-5c+9c$.

3. Solve the equation $3x-2 = 4-x$.

4. When I think of a number, double it and add three, I get 11. What number did I think of?

5. Solve the equation $x+2x-4 = 9$.

6. Simplify $2a+4-3+5a-a$.

EXERCISE 18n **1.** If $2x-9 = 2$, find x.

2. Simplify $6p+p-3p-4p$.

3. Find x if $4-x = 6-2x$.

4. If $3c+2 = c+2$, find c.

5. Peter had 14 marbles and lost x of them. John started with 8 marbles and gained x. The two boys then found that they each had the same number of marbles. Form an equation and find x.

6. Simplify $3b+4c-5c+c$.

EXERCISE 18p **1.** Find x if $9 = 3x-3$.

2. Simplify $2x+5x-8x$.

3. Solve the equation $3-2n = 5+3n$.

4. If $6x-4 = 2x+4$ find x.

5. Simplify $3a+2d-a+4c-3d+c$.

6. Solve the equation $2x+8-3x-6 = 4$.

19 VOLUME

In the science laboratory you may well have seen a container with a spout similar to the one shown in the diagram.

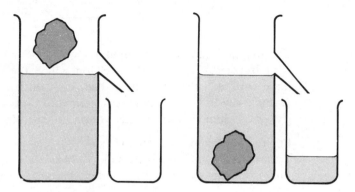

The container is filled with water to the level of the spout. Any solid which is put into the water will force a quantity of water into the measuring jug. The volume of this water will be equal to the volume of the solid. The volume of a solid is the amount of space it occupies.

CUBIC UNITS

As with area, we need a convenient unit for measuring volume. The most suitable unit is a cube.

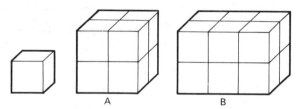

A   B

How many of the smallest cubes are needed to fill the same space as each of the solids A and B? Careful counting will show that 8 small cubes fill the same space as solid A and 12 small cubes fill the same space as solid B.

A cube with a side of 1 cm has a volume of one cubic centimetre which is written $1\,cm^3$.

Similarly a cube with a side of 1 mm has a volume of $1\,mm^3$ and a cube with a side of 1 m has a volume of $1\,m^3$.

VOLUME OF A CUBOID

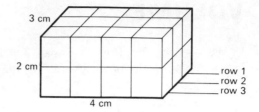

The diagram shows a rectangular block or cuboid measuring 4 cm by 3 cm by 2 cm. To cover the area on which the block stands we need three rows of cubes measuring 1 cm by 1 cm by 1 cm, with four cubes in each row, i.e. 12 cubes. A second layer of 12 cubes is needed to give the volume shown, so the volume of the block is 24 cubes.

But the volume of one cube is $1 \, cm^3$.
Therefore the volume of the solid is $24 \, cm^3$.
This is also given when we calculate length × breadth × height,

i.e.   the volume of the block $= 4 \times 3 \times 2 \, cm^3$

or   the volume of the block $=$ length × breadth × height

EXERCISE 19a

Find the volume of a cuboid measuring 12 cm by 10 cm by 5 cm.

Volume of cuboid $=$ length × breadth × height

$$= 12 \times 10 \times 5 \, cm^3$$

i.e.   Volume $= 600 \, cm^3$

Find the volume of each of the following cuboids:

	Length	Breadth	Height
1.	4 cm	4 cm	3 cm
2.	20 mm	10 mm	8 mm
3.	45 mm	20 mm	6 mm
4.	5 mm	4 mm	0.8 mm
5.	6.1 m	4 m	1.3 m
6.	3.5 cm	2.5 cm	1.2 cm

	Length	Breadth	Height
7.	4 m	3 m	2 m
8.	8 m	5 m	4 m
9.	8 cm	3 cm	$\frac{1}{2}$ cm
10.	12 cm	1.2 cm	0.5 cm
11.	4.5 m	1.2 m	0.8 m
12.	1.2 m	0.9 m	0.7 m

Find the volume of a cube with the given side:

6 cm

$$\text{Volume of cube} = \text{length} \times \text{breadth} \times \text{height}$$
$$= 6 \times 6 \times 6 \, \text{cm}^3$$

i.e.   Volume $= 216 \, \text{cm}^3$

13.	4 cm	**16.**	$\frac{1}{2}$ cm	**19.**	8 km
14.	5 cm	**17.**	2.5 cm	**20.**	$1\frac{1}{2}$ km
15.	2 m	**18.**	3 km	**21.**	3.4 m

EXERCISE 19b

Draw a cube of side 8 cm. How many cubes of side 2 cm would be needed to fill the same space?

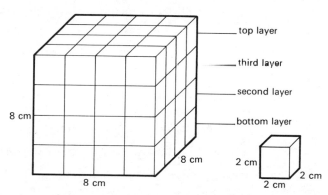

The bottom layer requires 4×4, i.e. 16 cubes of side 2 cm, and there are four layers altogether.

Therefore 64 cubes are required.

1. Draw a cube of side 4 cm. How many cubes of side 2 cm would be needed to fill the same space?

2. Draw a cuboid measuring 6 cm by 4 cm by 2 cm. How many cubes of side 2 cm would be needed to fill the same space?

3. Draw a cube of side 6 cm. How many cubes of side 3 cm would be needed to fill the same space?

4. Draw a cuboid measuring 8 cm by 6 cm by 2 cm. How many cubes of side 2 cm would be needed to fill the same space?

CHANGING UNITS OF VOLUME

Consider a cube of side 1 cm. If each edge is divided into 10 mm the cube can be divided into 10 layers, each layer with 10×10 cubes of side 1 mm,

i.e.
$$1 \, cm^3 = 10 \times 10 \times 10 \, mm^3$$

100 cubes, each with a volume of 1 mm³, in every one of these layers

Similarly, since 1 m = 100 cm

$$1 \text{ cubic metre} = 100 \times 100 \times 100 \, cm^3$$

i.e.
$$1 \, m^3 = 1\,000\,000 \, cm^3$$

EXERCISE 19c

Express $2.4 \, m^3$ in   a) cm^3 b) mm^3.

a) Since $1 \, m^3 = 100 \times 100 \times 100 \, cm^3$

$$2.4 \, m^3 = 2.4 \times 100 \times 100 \times 100 \, cm^3$$
$$= 2\,400\,000 \, cm^3$$

b) Since $1 \, m^3 = 1000 \times 1000 \times 1000 \, mm^3$

$$2.4 \, m^3 = 2.4 \times 1000 \times 1000 \times 1000 \, mm^3$$
$$= 2\,400\,000\,000 \, mm^3$$

Express in mm³:

1. 8 cm³ **3.** 6.2 cm³ **5.** 0.092 m³

2. 14 cm³ **4.** 0.43 cm³ **6.** 0.04 cm³

Express in cm³:

7. 3 m³ **9.** 0.42 m³ **11.** 22 mm³

8. 2.5 m³ **10.** 0.0063 m³ **12.** 731 mm³

CAPACITY

When we buy a bottle of milk or a can of engine oil we are not usually interested in the external measurements or volume of the container. What really concerns us is the *capacity* of the container, i.e. how much milk is inside the bottle, or how much engine oil is inside the can.

The most common unit of capacity in the metric system is the *litre*. (A litre is roughly equivalent to two bottles of milk.) A litre is much larger than a cubic centimetre but much smaller than a cubic metre. The relationship between these quantities is:

$$1000 \, cm^3 = 1 \text{ litre}$$

i.e. a litre is the volume of a cube of side 10 cm

and $\qquad 1000 \text{ litres} = 1 m^3$

When the amount of liquid is small, such as dosages for medicines, the millilitre (ml) is used. A millilitre is a thousandth part of a litre, i.e.

$$1000 \, ml = 1 \text{ litre} \quad or \quad 1 \, ml = 1 \, cm^3$$

EXERCISE 19d

Express 5.6 litres in cm³.

$$1 \text{ litre} = 1000 \, cm^3$$
$$\therefore \quad 5.6 \text{ litres} = 5.6 \times 1000 \, cm^3$$
$$= 5600 \, cm^3$$

Express in cm³:

1. 2.5 litres **4.** 0.0075 litres

2. 1.76 litres **5.** 35 litres

3. 0.54 litres **6.** 0.028 litres

Express in litres:

7. $7000 \, \text{cm}^3$ **9.** $24\,000 \, \text{cm}^3$

8. $4000 \, \text{cm}^3$ **10.** $600 \, \text{cm}^3$

Express in litres:

11. $5 \, \text{m}^3$ **13.** $4.6 \, \text{m}^3$

12. $12 \, \text{m}^3$ **14.** $0.067 \, \text{m}^3$

MIXED UNITS

Before we can find the volume of a cuboid, *all* measurements must be expressed in the same unit.

EXERCISE 19e

Find the volume of a cuboid measuring $2 \, \text{m}$ by $70 \, \text{cm}$ by $30 \, \text{cm}$.
Give your answer in   a) cm^3 b) m^3.

a) (All the measurements must be in centimetres so we first convert the $2 \, \text{m}$ into centimetres.)

$$\text{Length of cuboid} = 2 \, \text{m} = 2 \times 100 \, \text{cm} = 200 \, \text{cm}$$

$$\text{Volume of cuboid} = \text{length} \times \text{breadth} \times \text{height}$$

$$= 200 \times 70 \times 30 \, \text{cm}^3$$

$$= 420\,000 \, \text{cm}^3$$

b) (We convert all the measurements to metres before finding the volume.)

$$\text{Breadth of cuboid} = 70 \, \text{cm} = \frac{70}{100} \, \text{m} = 0.7 \, \text{m}$$

$$\text{Height of cuboid} = 30 \, \text{cm} = \frac{30}{100} \, \text{m} = 0.3 \, \text{m}$$

$$\therefore \quad \text{Volume of cuboid} = 2 \times 0.7 \times 0.3 \, \text{m}^3$$

$$= 0.42 \, \text{m}^3$$

Find the volumes of the following cuboids, giving your answers in the units stated in brackets:

	Length	Breadth	Height	
1.	$50 \, \text{mm}$	$30 \, \text{mm}$	$20 \, \text{mm}$	(cm^3)
2.	$400 \, \text{cm}$	$100 \, \text{cm}$	$50 \, \text{cm}$	(m^3)

	Length	Breadth	Height	
3.	1 m	4 cm	2 cm	(cm³)
4.	15 cm	80 mm	50 mm	(cm³)
5.	6 cm	12 mm	8 mm	(mm³)
6.	2 m	50 cm	40 mm	(cm³)
7.	4 cm	35 mm	2 cm	(cm³)
8.	20 m	80 cm	50 cm	(m³)
9.	3.5 cm	25 mm	20 mm	(cm³)
10.	$\frac{1}{2}$ m	45 mm	8 mm	(cm³)

PROBLEMS INVOLVING CUBOIDS

EXERCISE 19f

Find the volume of a rectangular block of wood measuring 8 cm by 6 cm which is 2 m long. Give your answer in cubic centimetres.

Working in centimetres:

$$\text{Length of block} = 2\,\text{m} = 2 \times 100\,\text{cm} = 200\,\text{cm}$$

$$\text{Volume of block} = \text{length} \times \text{breadth} \times \text{height}$$

$$= 200 \times 8 \times 6\,\text{cm}^3$$

$$= 9600\,\text{cm}^3$$

A rectangular metal water tank is 3 m long, 2.5 m wide and 80 cm deep.
Find its capacity in a) m³ b) litres.

a) Working in metres:

$$\text{Depth of tank} = 80\,\text{cm} = \frac{80}{100}\,\text{m} = 0.8\,\text{m}$$

$$\text{Capacity of tank} = \text{length} \times \text{breadth} \times \text{height}$$

$$= 3 \times 2.5 \times 0.8\,\text{m}^3$$

$$= 6\,\text{m}^3$$

b) $$1\,\text{m}^3 = 1000\,\text{litres}$$

$$\text{Capacity of tank} = 6 \times 1000\,\text{litres}$$

$$= 6000\,\text{litres}$$

1. Find the volume of air in a room measuring 4 m by 5 m which is 3 m high.

2. Find the volume, in cm³, of a concrete block measuring 36 cm by 18 cm by 12 cm.

3. Find the volume of a school hall which is 30 m long and 24 m wide if the ceiling is 9 m high.

4. An electric light bulb is sold in a box measuring 10 cm by 6 cm by 6 cm. If the shopkeeper receives them in a carton measuring 50 cm by 30 cm by 30 cm, how many bulbs would be packed in a carton?

5. A classroom is 10 m long, 8 m wide and 3 m high. How many pupils should it be used for if each pupil requires 5 m³ of air space?

6. How many cubic metres of water are required to fill a rectangular swimming bath 15 m long and 10 m wide which is 2 m deep throughout? How many litres is this?

7. How many rectangular packets, measuring 8 cm by 6 cm by 4 cm, may be packed in a rectangular cardboard box measuring 30 cm by 24 cm by 16 cm?

8. A water storage tank is 3 m long, 2 m wide and $1\frac{1}{2}$ m deep. How many litres of water will it hold?

9. How many lead cubes of side 2 cm could be made from a lead cube of side 8 cm?

10. How many lead cubes of side 5 mm could be made from a rectangular block of lead measuring 10 cm by 5 cm by 4 cm?

MIXED EXERCISES

EXERCISE 19g 1. Express 3.2 m³ in   a) cm³   b) mm³.

2. Express 1.6 litres in cm³.

3. Find the volume of a cube of side 4 cm.

4. Find the volume, in cm³, of a cuboid measuring 2 m by 25 cm by 10 cm.

5. Find the volume, in mm³, of a cuboid measuring 5 cm by 3 cm by 9 mm.

EXERCISE 19h **1.** Express $8\,cm^3$ in   a) mm^3 b) m^3.

2. Express $3500\,cm^3$ in litres.

3. Find the volume of a cuboid measuring 10 cm by 5 cm by 6 cm.

4. Find, in cm^3, the volume of a cube of side 8 mm.

5. Find the volume, in cm^3, of a cuboid measuring 50 cm by 1.2 m by 20 cm.

EXERCISE 19i **1.** Express $0.009\,m^3$ in   a) cm^3 b) mm^3.

2. Express 0.44 litres in cm^3.

3. Find the volume of a cube of side 6 cm.

4. Find the volume of a cuboid measuring 12 cm by 6 cm by 4 cm.

5. Find the capacity, in litres, of a rectangular tank measuring 2 m by 1.5 m by 80 cm.

EXERCISE 19j **1.** Express $900\,cm^3$ in m^3.

2. Express $10\,800\,cm^3$ in litres.

3. Express $0.075\,m^3$ in litres.

4. Find, in cm^3, the volume of a cube of side 20 mm.

5. Find, in m^3, the volume of a cuboid measuring 150 cm by 100 cm by 80 cm.

20 VECTORS

If you arranged to meet your friend 3 km from your home, this information would not be enough to ensure that you both went to the same place. You would also need to know which way to go.

Two pieces of information are required to describe where one place is in relation to another: the distance and the direction. Quantities which have both *size* (magnitude) and *direction* are called *vectors*.

A quantity which has magnitude but not direction is called a *scalar*. For example, the amount of money in your pocket or the number of pupils in your school are scalar quantities.

EXERCISE 20a State whether the following sentences refer to vector or scalar quantities:

1. There are 24 pupils in my class.

2. To get to school I walk $\frac{1}{2}$ km due north.

3. There are 11 players in a cricket team.

4. John walked at 6 km per hour.

5. The vertical cliff face is 50 m high.

6. Give other examples of
   a) vector quantities   b) scalar quantities.

REPRESENTING VECTORS

We can represent a vector by a straight line and indicate its direction with an arrow. For example

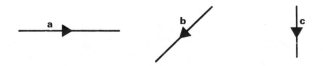

We use **a**, **b**, **c**, ... to name the vectors.

When writing by hand it is difficult to write **a**, which is in heavy type, so we use $\underline{a}$.

278

In the diagram below, the movement along **a** corresponds to 4 across and 2 up and we can write

$$\mathbf{a} \quad = \quad \begin{pmatrix} 4 \\ 2 \end{pmatrix}$$

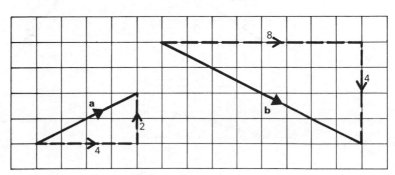

The vector **b** can be described as 8 across and 4 down. As with coordinates, which we looked at in Chapter 16, we use negative numbers to indicate movement down or movement to the left.

Therefore

$$\mathbf{b} \quad = \quad \begin{pmatrix} 8 \\ -4 \end{pmatrix}$$

Notice that the top number represents movement across and that the bottom number represents movement up or down.

EXERCISE 20b

Write the following vectors in the form $\begin{pmatrix} p \\ q \end{pmatrix}$:

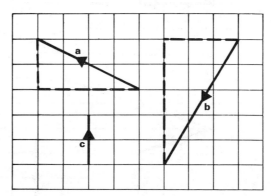

$$\mathbf{a} = \begin{pmatrix} -4 \\ 2 \end{pmatrix}, \qquad \mathbf{b} = \begin{pmatrix} -3 \\ -5 \end{pmatrix}, \qquad \mathbf{c} = \begin{pmatrix} 0 \\ 2 \end{pmatrix}$$

Write the following vectors in the form $\begin{pmatrix} p \\ q \end{pmatrix}$:

1.

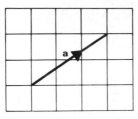

5.

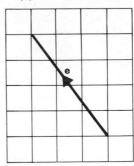

2.

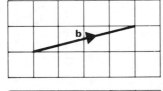

3.

6.

4.

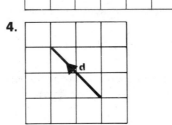

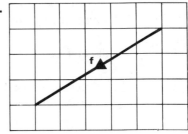

7.

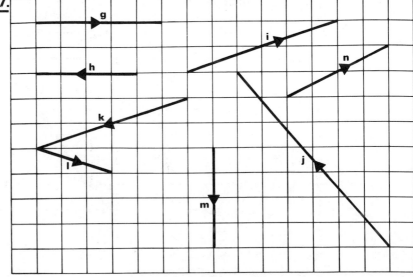

On squared paper draw the following vectors. Label each vector with its letter and an arrow:

8. $\mathbf{a} = \begin{pmatrix} 3 \\ 5 \end{pmatrix}$

9. $\mathbf{b} = \begin{pmatrix} -4 \\ -3 \end{pmatrix}$

10. $\mathbf{c} = \begin{pmatrix} 2 \\ -4 \end{pmatrix}$

11. $\mathbf{d} = \begin{pmatrix} 6 \\ -12 \end{pmatrix}$

12. $\mathbf{e} = \begin{pmatrix} -4 \\ 3 \end{pmatrix}$

13. $\mathbf{f} = \begin{pmatrix} -2 \\ 5 \end{pmatrix}$

14. $\mathbf{g} = \begin{pmatrix} 6 \\ 10 \end{pmatrix}$

15. $\mathbf{h} = \begin{pmatrix} -1 \\ -5 \end{pmatrix}$

16. $\mathbf{i} = \begin{pmatrix} -6 \\ 2 \end{pmatrix}$

17. $\mathbf{j} = \begin{pmatrix} 5 \\ -3 \end{pmatrix}$

18. What do you notice about the vectors in questions 8 and 14, and in questions 10 and 11?

EXERCISE 20c In each of the following questions you are given a vector followed by the coordinates of its starting point. Find the coordinates of its other end:

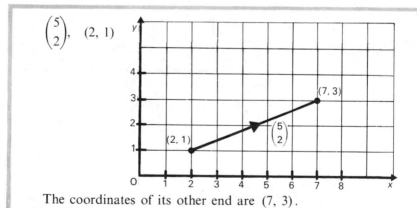

$\begin{pmatrix} 5 \\ 2 \end{pmatrix}$,   (2, 1)

The coordinates of its other end are (7, 3).

1. $\begin{pmatrix} 3 \\ 3 \end{pmatrix}$,   (4, 1)

2. $\begin{pmatrix} 3 \\ 1 \end{pmatrix}$,   (−2, −3)

3. $\begin{pmatrix} -6 \\ 2 \end{pmatrix}$,   (3, 5)

4. $\begin{pmatrix} -4 \\ -3 \end{pmatrix}$,   (5, −2)

5. $\begin{pmatrix} 5 \\ 2 \end{pmatrix}$,   (3, −1)

6. $\begin{pmatrix} 4 \\ -2 \end{pmatrix}$,   (4, 2)

7. $\begin{pmatrix} -3 \\ 4 \end{pmatrix}$, (2, −4)   **10.** $\begin{pmatrix} 5 \\ -3 \end{pmatrix}$, (2, −1)

8. $\begin{pmatrix} -6 \\ -6 \end{pmatrix}$, (−3, −2)   **11.** $\begin{pmatrix} -5 \\ 2 \end{pmatrix}$, (−4, −3)

9. $\begin{pmatrix} 4 \\ 3 \end{pmatrix}$, (−2, −3)   **12.** $\begin{pmatrix} -4 \\ -2 \end{pmatrix}$, (−3, −1)

In each of the following questions a vector is given followed by the coordinates of its other end. Find the coordinates of its starting point:

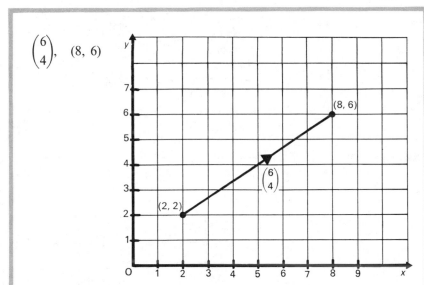

$\begin{pmatrix} 6 \\ 4 \end{pmatrix}$, (8, 6)

The coordinates of the vector's starting point are (2, 2).

13. $\begin{pmatrix} 10 \\ 2 \end{pmatrix}$, (4, 1)   **18.** $\begin{pmatrix} -6 \\ -3 \end{pmatrix}$, (−5, 2)

14. $\begin{pmatrix} 5 \\ -1 \end{pmatrix}$, (3, −4)   **19.** $\begin{pmatrix} 4 \\ -2 \end{pmatrix}$, (−3, 2)

15. $\begin{pmatrix} -5 \\ -2 \end{pmatrix}$, (−2, −4)   **20.** $\begin{pmatrix} -2 \\ 6 \end{pmatrix}$, (−3, −4)

16. $\begin{pmatrix} 8 \\ 6 \end{pmatrix}$, (6, 3)   **21.** $\begin{pmatrix} 1 \\ 4 \end{pmatrix}$, (−5, −2)

17. $\begin{pmatrix} -3 \\ 4 \end{pmatrix}$, (−2, 1)   **22.** $\begin{pmatrix} 2 \\ -3 \end{pmatrix}$, (1, 7)

CAPITAL LETTER NOTATION

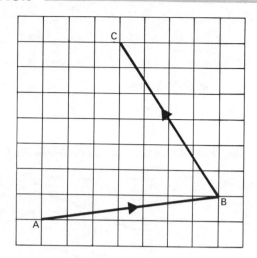

In the diagram A and B are two points.

We can denote the vector from A to B as $\overrightarrow{AB}$ where $\overrightarrow{AB} = \begin{pmatrix} 7 \\ 1 \end{pmatrix}$.

Similarly $\overrightarrow{BC} = \begin{pmatrix} -4 \\ 6 \end{pmatrix}$.

EXERCISE 20d

Write down the vector $\overrightarrow{AB}$ where A is (2, 4), B is (3, 9).

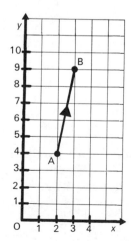

Vector $\overrightarrow{AB}$ is $\begin{pmatrix} 1 \\ 5 \end{pmatrix}$.

Write down the vector $\overrightarrow{AB}$ where:

1. A is (1, 4), B is (7, 6)   **6.** A is (3, 0), B is (5, −2)

2. A is (−3, 4), B is (2, 3)   **7.** A is (6, 3), B is (4, 1)

3. A is (7, 3), B is (1, 2)   **8.** A is (−1, −3), B is (−5, −8)

4. A is (−1, 4), B is (5, 9)   **9.** A is (2, 6), B is (2, −6)

5. A is (2, 1), B is (−3, −5)   **10.** A is (2, −3), B is (4, 5)

EQUAL VECTORS, PARALLEL VECTORS AND NEGATIVE VECTORS

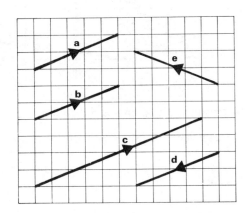

In the diagram, we can see that

$$\mathbf{a} = \begin{pmatrix} 5 \\ 2 \end{pmatrix} \qquad \text{and} \qquad \mathbf{b} = \begin{pmatrix} 5 \\ 2 \end{pmatrix}$$

so we say that $\mathbf{a} = \mathbf{b}$

The lines representing **a** and **b** are parallel and equal in length.

Now $\mathbf{c} = \begin{pmatrix} 10 \\ 4 \end{pmatrix} \qquad \text{and} \qquad \mathbf{a} = \begin{pmatrix} 5 \\ 2 \end{pmatrix}$

so $\mathbf{c} = 2\mathbf{a}$

This time **a** and **c** are parallel but **c** is twice the size of **a**.

If we look at **a** and **d** we have

$$\mathbf{d} = \begin{pmatrix} -5 \\ -2 \end{pmatrix} \qquad \text{and} \qquad \mathbf{a} = \begin{pmatrix} 5 \\ 2 \end{pmatrix}$$

Although **d** and **a** are parallel and the same size, they are in opposite directions,

so we say that $\mathbf{d} = -\mathbf{a}$

Now
$$e = \begin{pmatrix} -5 \\ 2 \end{pmatrix}$$

and we can see from the diagram that although **e** is the same size as **a**, **b** and **d**, it is *not* parallel to them or to **c**. So **e** is not equal to any of the other vectors.

EXERCISE 20e 1. Write the vectors **a**, **b**, **c**, **d** and **e** in the form $\begin{pmatrix} p \\ q \end{pmatrix}$.

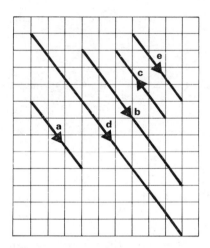

What is the relationship between

a) **a** and **b**   b) **a** and **c**   c) **a** and **d**

d) **a** and **e**   e) **b** and **e**   f) **d** and **c**?

2.

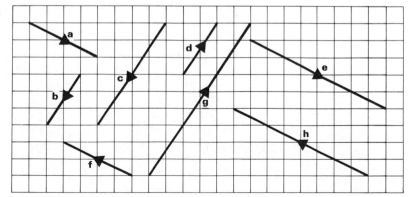

Find as many relationships as you can between the vectors in the diagram.

3. If $\mathbf{a} = \begin{pmatrix} 4 \\ 6 \end{pmatrix}$ draw diagrams to represent **a**, 2**a**, −**a**, $\frac{1}{2}$**a**.

   Now write the vectors 2**a**, −**a**, $\frac{1}{2}$**a** in the form $\begin{pmatrix} p \\ q \end{pmatrix}$.

4. If $\mathbf{b} = \begin{pmatrix} -2 \\ 4 \end{pmatrix}$ draw diagrams to represent **b**, −**b**, 2**b**, −2**b**.

   Now write the vectors −**b**, 2**b**, −2**b** in the form $\begin{pmatrix} p \\ q \end{pmatrix}$.

5. If $\mathbf{c} = \begin{pmatrix} 5 \\ -4 \end{pmatrix}$ draw diagrams to represent **c**, 2**c**, −**c**, 3**c**.

   Now write the vectors 2**c**, −**c**, 3**c** in the form $\begin{pmatrix} p \\ q \end{pmatrix}$.

6. If $\mathbf{d} = \begin{pmatrix} -3 \\ -6 \end{pmatrix}$ draw diagrams to represent **d**, −**d**, 2**d**, −2**d**.

   Now write the vectors −**d**, 2**d**, −2**d** in the form $\begin{pmatrix} p \\ q \end{pmatrix}$.

7. If $\mathbf{e} = \begin{pmatrix} 5 \\ 1 \end{pmatrix}$ write, in the form $\begin{pmatrix} p \\ q \end{pmatrix}$. the vectors 2**e**, −**e**, 3**e** and −4**e**.

8. If $\mathbf{f} = \begin{pmatrix} -2 \\ 0 \end{pmatrix}$ write, in the form $\begin{pmatrix} p \\ q \end{pmatrix}$, the vectors 3**f**, −2**f**, 5**f** and −4**f**.

9. If $\mathbf{g} = \begin{pmatrix} 6 \\ -4 \end{pmatrix}$ write, in the form $\begin{pmatrix} p \\ q \end{pmatrix}$, the vectors −**g**, 3**g**, $\frac{1}{2}$**g**, and −2**g**.

10. If $\mathbf{h} = \begin{pmatrix} -6 \\ -20 \end{pmatrix}$ write, in the form $\begin{pmatrix} p \\ q \end{pmatrix}$, the vectors 3**h**, −4**h**, $\frac{1}{2}$**h**, and −5**h**.

ADDITION OF VECTORS

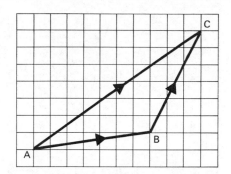

The displacement from A to B followed by the displacement from B to C is equivalent to the displacement from A to C, so we write

$$\overrightarrow{AB} + \overrightarrow{BC} = \overrightarrow{AC}$$

or

$$\begin{pmatrix} 7 \\ 1 \end{pmatrix} + \begin{pmatrix} 3 \\ 6 \end{pmatrix} = \begin{pmatrix} 10 \\ 7 \end{pmatrix}$$

To add two vectors we add the corresponding numbers in the ordered pairs that represent them.

EXERCISE 20f

If $\mathbf{a} = \begin{pmatrix} 2 \\ 4 \end{pmatrix}$ and $\mathbf{b} = \begin{pmatrix} 6 \\ -8 \end{pmatrix}$ find $\mathbf{a}+\mathbf{b}$ and illustrate on a diagram.

$$\mathbf{a}+\mathbf{b} = \begin{pmatrix} 2 \\ 4 \end{pmatrix} + \begin{pmatrix} 6 \\ -8 \end{pmatrix}$$

$$= \begin{pmatrix} 8 \\ -4 \end{pmatrix}$$

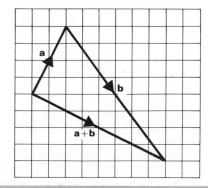

In questions 1 to 10 find $\mathbf{a}+\mathbf{b}$ and illustrate on a diagram:

1. $\mathbf{a} = \begin{pmatrix} 6 \\ 4 \end{pmatrix}$, $\mathbf{b} = \begin{pmatrix} 1 \\ -5 \end{pmatrix}$ 6. $\mathbf{a} = \begin{pmatrix} 4 \\ 0 \end{pmatrix}$, $\mathbf{b} = \begin{pmatrix} 0 \\ 3 \end{pmatrix}$

2. $\mathbf{a} = \begin{pmatrix} -3 \\ 4 \end{pmatrix}$, $\mathbf{b} = \begin{pmatrix} -5 \\ -2 \end{pmatrix}$ 7. $\mathbf{a} = \begin{pmatrix} -2 \\ -4 \end{pmatrix}$, $\mathbf{b} = \begin{pmatrix} -4 \\ -2 \end{pmatrix}$

3. $\mathbf{a} = \begin{pmatrix} 5 \\ 0 \end{pmatrix}$, $\mathbf{b} = \begin{pmatrix} 2 \\ -4 \end{pmatrix}$ 8. $\mathbf{a} = \begin{pmatrix} 5 \\ 2 \end{pmatrix}$, $\mathbf{b} = \begin{pmatrix} -2 \\ 4 \end{pmatrix}$

4. $\mathbf{a} = \begin{pmatrix} -4 \\ 3 \end{pmatrix}$, $\mathbf{b} = \begin{pmatrix} 6 \\ 3 \end{pmatrix}$ 9. $\mathbf{a} = \begin{pmatrix} 2 \\ 6 \end{pmatrix}$, $\mathbf{b} = \begin{pmatrix} 5 \\ 2 \end{pmatrix}$

5. $\mathbf{a} = \begin{pmatrix} 5 \\ 3 \end{pmatrix}$, $\mathbf{b} = \begin{pmatrix} 5 \\ -3 \end{pmatrix}$ 10. $\mathbf{a} = \begin{pmatrix} 5 \\ 1 \end{pmatrix}$, $\mathbf{b} = \begin{pmatrix} 1 \\ -5 \end{pmatrix}$

In questions 11 to 20 find the following vectors:

11. $\begin{pmatrix} 2 \\ 6 \end{pmatrix} + \begin{pmatrix} 4 \\ 3 \end{pmatrix}$ 16. $\begin{pmatrix} 1 \\ 4 \end{pmatrix} + \begin{pmatrix} -3 \\ 6 \end{pmatrix}$

12. $\begin{pmatrix} 5 \\ 5 \end{pmatrix} + \begin{pmatrix} 2 \\ 6 \end{pmatrix}$ 17. $\begin{pmatrix} 2 \\ 1 \end{pmatrix} + \begin{pmatrix} -4 \\ -5 \end{pmatrix}$

13. $\begin{pmatrix} 4 \\ 9 \end{pmatrix} + \begin{pmatrix} 3 \\ 1 \end{pmatrix}$ 18. $\begin{pmatrix} -7 \\ 2 \end{pmatrix} + \begin{pmatrix} 2 \\ -4 \end{pmatrix}$

14. $\begin{pmatrix} 6 \\ 2 \end{pmatrix} + \begin{pmatrix} 4 \\ -2 \end{pmatrix}$ 19. $\begin{pmatrix} -5 \\ 9 \end{pmatrix} + \begin{pmatrix} -3 \\ -4 \end{pmatrix}$

15. $\begin{pmatrix} 5 \\ 9 \end{pmatrix} + \begin{pmatrix} -6 \\ 2 \end{pmatrix}$ 20. $\begin{pmatrix} 2 \\ 4 \end{pmatrix} + \begin{pmatrix} -2 \\ -4 \end{pmatrix}$

ORDER OF ADDITION

If $\mathbf{a} = \begin{pmatrix} 4 \\ 2 \end{pmatrix}$ and $\mathbf{b} = \begin{pmatrix} 2 \\ 3 \end{pmatrix}$

$$\mathbf{a} + \mathbf{b} = \begin{pmatrix} 4 \\ 2 \end{pmatrix} + \begin{pmatrix} 2 \\ 3 \end{pmatrix} = \begin{pmatrix} 6 \\ 5 \end{pmatrix}$$

and

$$\mathbf{b} + \mathbf{a} = \begin{pmatrix} 2 \\ 3 \end{pmatrix} + \begin{pmatrix} 4 \\ 2 \end{pmatrix} = \begin{pmatrix} 6 \\ 5 \end{pmatrix}$$

i.e. $\mathbf{a} + \mathbf{b} = \mathbf{b} + \mathbf{a}$

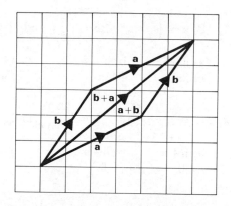

i.e.   the order in which you do the addition does not matter.

ADDITION OF MORE THAN TWO VECTORS

If $\mathbf{a} = \begin{pmatrix} 4 \\ 1 \end{pmatrix}$, $\mathbf{b} = \begin{pmatrix} 4 \\ 4 \end{pmatrix}$ and $\mathbf{c} = \begin{pmatrix} -2 \\ 4 \end{pmatrix}$ we can find $\mathbf{a} + \mathbf{b} + \mathbf{c}$ by adding the corresponding numbers in the ordered pairs.

$$\mathbf{a} + \mathbf{b} + \mathbf{c} = \begin{pmatrix} 4 \\ 1 \end{pmatrix} + \begin{pmatrix} 4 \\ 4 \end{pmatrix} + \begin{pmatrix} -2 \\ 4 \end{pmatrix} = \begin{pmatrix} 6 \\ 9 \end{pmatrix}$$

Again the order of addition does not matter, as you can see from the diagrams below:

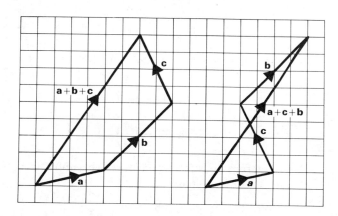

i.e. $\mathbf{a} + \mathbf{b} + \mathbf{c} = \mathbf{a} + \mathbf{c} + \mathbf{b}$

There are other possible orders in which we could add $\mathbf{a}$, $\mathbf{b}$ and $\mathbf{c}$.

EXERCISE 20g 1. If $\mathbf{a} = \begin{pmatrix} 2 \\ 3 \end{pmatrix}$, $\mathbf{b} = \begin{pmatrix} 5 \\ 2 \end{pmatrix}$ and $\mathbf{c} = \begin{pmatrix} 3 \\ 4 \end{pmatrix}$ find:

a) $\mathbf{a}+\mathbf{b}$ b) $\mathbf{b}+\mathbf{a}$ c) $\mathbf{b}+\mathbf{c}$ d) $\mathbf{c}+\mathbf{b}$
e) $2\mathbf{a}$ f) $3\mathbf{a}$ g) $\mathbf{a}+\mathbf{b}+\mathbf{c}$ h) $\mathbf{c}+\mathbf{b}+\mathbf{a}$

2. If $\mathbf{a} = \begin{pmatrix} 5 \\ -2 \end{pmatrix}$, $\mathbf{b} = \begin{pmatrix} -2 \\ 4 \end{pmatrix}$ and $\mathbf{c} = \begin{pmatrix} -5 \\ -3 \end{pmatrix}$ find:

a) $\mathbf{a}+\mathbf{b}$ b) $\mathbf{b}+\mathbf{a}$ c) $\mathbf{a}+\mathbf{c}$ d) $\mathbf{c}+\mathbf{a}$
e) $3\mathbf{b}$ f) $4\mathbf{c}$

3. If $\mathbf{a} = \begin{pmatrix} 3 \\ 2 \end{pmatrix}$, $\mathbf{b} = \begin{pmatrix} -3 \\ 2 \end{pmatrix}$ and $\mathbf{c} = \begin{pmatrix} 5 \\ 6 \end{pmatrix}$ find:

a) $\mathbf{a}+\mathbf{b}+\mathbf{c}$ b) $2\mathbf{a}+\mathbf{b}+3\mathbf{c}$ c) $\mathbf{a}+2\mathbf{b}+3\mathbf{c}$

4. If $\mathbf{a} = \begin{pmatrix} -6 \\ -2 \end{pmatrix}$, $\mathbf{b} = \begin{pmatrix} 4 \\ -3 \end{pmatrix}$ and $\mathbf{c} = \begin{pmatrix} -5 \\ 3 \end{pmatrix}$ find:

a) $2\mathbf{a}+2\mathbf{b}+3\mathbf{c}$ b) $\mathbf{a}+5\mathbf{b}+2\mathbf{c}$

SUBTRACTION OF VECTORS

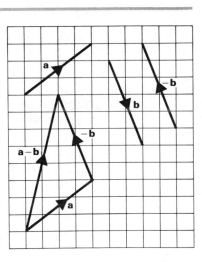

If $\mathbf{a} = \begin{pmatrix} 4 \\ 3 \end{pmatrix}$ and $\mathbf{b} = \begin{pmatrix} 2 \\ -5 \end{pmatrix}$

we know that $\mathbf{a}-\mathbf{b} = \mathbf{a}+(-\mathbf{b})$

$$= \begin{pmatrix} 4 \\ 3 \end{pmatrix} + \begin{pmatrix} -2 \\ 5 \end{pmatrix}$$

$$= \begin{pmatrix} 2 \\ 8 \end{pmatrix}$$

But this result is also given by

$$\begin{pmatrix} 4 \\ 3 \end{pmatrix} - \begin{pmatrix} 2 \\ -5 \end{pmatrix}$$

$\therefore$ $\mathbf{a}-\mathbf{b} = \begin{pmatrix} 4 \\ 3 \end{pmatrix} - \begin{pmatrix} 2 \\ -5 \end{pmatrix}$

Therefore to subtract vectors we subtract the corresponding numbers in the ordered pairs.

Note that $\mathbf{b}-\mathbf{a} = \begin{pmatrix} 2 \\ -5 \end{pmatrix} - \begin{pmatrix} 4 \\ 3 \end{pmatrix} = \begin{pmatrix} -2 \\ -8 \end{pmatrix}$

and this is *not* the same as $\mathbf{a}-\mathbf{b}$.
So in subtraction, the order *does* matter.

EXERCISE 20h In questions 1 to 4, find **a**−**b** and draw diagrams to represent **a**, **b** and **a**−**b** :

$$\mathbf{a} = \begin{pmatrix} 3 \\ 6 \end{pmatrix}, \mathbf{b} = \begin{pmatrix} -2 \\ 3 \end{pmatrix}$$

$$\mathbf{a} - \mathbf{b} = \begin{pmatrix} 3 \\ 6 \end{pmatrix} - \begin{pmatrix} -2 \\ 3 \end{pmatrix} = \begin{pmatrix} 5 \\ 3 \end{pmatrix}$$

1. $\mathbf{a} = \begin{pmatrix} 2 \\ 5 \end{pmatrix}, \mathbf{b} = \begin{pmatrix} -3 \\ 2 \end{pmatrix}$

3. $\mathbf{a} = \begin{pmatrix} 6 \\ 3 \end{pmatrix}, \mathbf{b} = \begin{pmatrix} 4 \\ -1 \end{pmatrix}$

2. $\mathbf{a} = \begin{pmatrix} -3 \\ 4 \end{pmatrix}, \mathbf{b} = \begin{pmatrix} -3 \\ -2 \end{pmatrix}$

4. $\mathbf{a} = \begin{pmatrix} -4 \\ -3 \end{pmatrix}, \mathbf{b} = \begin{pmatrix} 1 \\ -4 \end{pmatrix}$

Find the following vectors:

5. $\begin{pmatrix} 10 \\ 3 \end{pmatrix} - \begin{pmatrix} 8 \\ 2 \end{pmatrix}$

9. $\begin{pmatrix} 6 \\ 2 \end{pmatrix} - \begin{pmatrix} 9 \\ 4 \end{pmatrix}$

6. $\begin{pmatrix} 6 \\ 4 \end{pmatrix} - \begin{pmatrix} 3 \\ 2 \end{pmatrix}$

10. $\begin{pmatrix} 3 \\ 5 \end{pmatrix} - \begin{pmatrix} 10 \\ 2 \end{pmatrix}$

7. $\begin{pmatrix} 9 \\ 12 \end{pmatrix} - \begin{pmatrix} -2 \\ 3 \end{pmatrix}$

11. $\begin{pmatrix} -2 \\ 7 \end{pmatrix} - \begin{pmatrix} -4 \\ 3 \end{pmatrix}$

8. $\begin{pmatrix} 4 \\ 5 \end{pmatrix} - \begin{pmatrix} -1 \\ -3 \end{pmatrix}$

12. $\begin{pmatrix} -5 \\ -2 \end{pmatrix} - \begin{pmatrix} -6 \\ -1 \end{pmatrix}$

13. $\begin{pmatrix} 4 \\ 3 \end{pmatrix} + \begin{pmatrix} 2 \\ 3 \end{pmatrix} + \begin{pmatrix} -1 \\ 4 \end{pmatrix}$

16. $\begin{pmatrix} 3 \\ 1 \end{pmatrix} - \begin{pmatrix} 4 \\ 2 \end{pmatrix} - \begin{pmatrix} -3 \\ -4 \end{pmatrix}$

14. $\begin{pmatrix} 5 \\ 2 \end{pmatrix} + \begin{pmatrix} -3 \\ -4 \end{pmatrix} + \begin{pmatrix} 2 \\ -3 \end{pmatrix}$

17. $\begin{pmatrix} 3 \\ 2 \end{pmatrix} + \begin{pmatrix} -4 \\ 3 \end{pmatrix} - \begin{pmatrix} 2 \\ -6 \end{pmatrix}$

15. $\begin{pmatrix} 7 \\ 6 \end{pmatrix} - \begin{pmatrix} 2 \\ 3 \end{pmatrix} - \begin{pmatrix} 1 \\ 4 \end{pmatrix}$

18. $\begin{pmatrix} -5 \\ -2 \end{pmatrix} - \begin{pmatrix} 4 \\ -3 \end{pmatrix} + \begin{pmatrix} -2 \\ 6 \end{pmatrix}$

19. If $\mathbf{a} = \begin{pmatrix} 6 \\ 2 \end{pmatrix}$ and $\mathbf{b} = \begin{pmatrix} 5 \\ 0 \end{pmatrix}$ find:

   a) $\mathbf{a} - \mathbf{b}$ b) $\mathbf{b} - \mathbf{a}$

20. If $\mathbf{a} = \begin{pmatrix} -4 \\ 2 \end{pmatrix}$, $\mathbf{b} = \begin{pmatrix} 2 \\ 6 \end{pmatrix}$ and $\mathbf{c} = \begin{pmatrix} 5 \\ 9 \end{pmatrix}$ find:

   a) $\mathbf{a} - \mathbf{b}$ b) $\mathbf{b} - \mathbf{c}$ c) $\mathbf{c} - \mathbf{b}$

21. If $\mathbf{a} = \begin{pmatrix} 5 \\ 4 \end{pmatrix}$, $\mathbf{b} = \begin{pmatrix} 2 \\ 6 \end{pmatrix}$ and $\mathbf{c} = \begin{pmatrix} -3 \\ -1 \end{pmatrix}$ find:

   a) $2\mathbf{a} - \mathbf{b}$ b) $3\mathbf{b} - \mathbf{c}$ c) $2\mathbf{a} - 5\mathbf{b}$

   d) $\mathbf{a} + \mathbf{b} - \mathbf{c}$ e) $\mathbf{b} - \mathbf{c} - \mathbf{a}$

22. If $\mathbf{a} = \begin{pmatrix} -2 \\ 6 \end{pmatrix}$, $\mathbf{b} = \begin{pmatrix} 4 \\ 1 \end{pmatrix}$ and $\mathbf{c} = \begin{pmatrix} 3 \\ -5 \end{pmatrix}$ find:

   a) $2\mathbf{a} + \mathbf{b} - \mathbf{c}$ b) $\mathbf{a} - \mathbf{b} + \mathbf{c}$ c) $2\mathbf{a} + \mathbf{b} + \mathbf{c}$

   d) $-3\mathbf{b} + 4\mathbf{c}$ e) $2\mathbf{b} - 4\mathbf{c}$

23. If $\mathbf{a} = \begin{pmatrix} -3 \\ -2 \end{pmatrix}$, $\mathbf{b} = \begin{pmatrix} 2 \\ -5 \end{pmatrix}$ and $\mathbf{c} = \begin{pmatrix} -4 \\ -2 \end{pmatrix}$ find:

   a) $\mathbf{a} + 2\mathbf{b} - \mathbf{c}$ b) $3\mathbf{a} - 4\mathbf{b}$ c) $4\mathbf{b} - 3\mathbf{c}$

21 MORE ALGEBRA

BRACKETS

Sometimes brackets are used to hold two quantities together. For instance, if we wish to multiply both x and 3 by 4 we write $4(x+3)$. The multiplication sign is invisible just as it is in $5x$, which means $5 \times x$.

$4(x+3)$ means "four times everything in the brackets" so we have $4 \times x$ and 4×3, and we write $4(x+3) = 4x+12$.

EXERCISE 21a Multiply out the following brackets:

$3(4x+2)$

$$3(4x+2) = 12x+6$$

$2(x-1)$

$$2(x-1) = 2x-2$$

1.	$2(x+1)$	**5.**	$2(4+5x)$	**9.**	$3(6-4x)$
2.	$3(3x-2)$	**6.**	$2(6+5a)$	**10.**	$5(x-1)$
3.	$5(x+6)$	**7.**	$5(a+b)$	**11.**	$7(2-x)$
4.	$4(3x-3)$	**8.**	$4(4x-3)$	**12.**	$8(3-2x)$

To simplify an expression containing brackets we first multiply out the brackets and then collect like terms.

EXERCISE 21b Simplify the following expressions:

$6x+3(x-2)$

$$6x+3(x-2) = 6x+3x-6$$
$$= 9x-6$$

$2+(3x-7)$

$$2+(3x-7) \qquad \text{(This means } 2+1(3x-7))$$
$$= 2+3x-7$$
$$= 3x-5$$

Simplify the following expressions:

1. $2x+4(x+1)$ **6.** $3x+3(x-5)$

2. $3+5(2x+3)$ **7.** $2(x+4)+3(x+5)$

3. $3(x+1)+4$ **8.** $6(2x-3)+5(x-1)$

4. $6(2x-3)+2x$ **9.** $3x+(2x+5)$

5. $7+2(2x+5)$ **10.** $4+(3x-1)$

$4x-2(x+3)$

(In the expression $4x-2(x+3)$ we see that we are required to take away 2 x's and 2 threes (which is 6).)

$$4x-2(x+3) = 4x-2x-6$$
$$= 2x-6$$

$5-(x+4)$

$$5-(x+4) = 5-x-4$$
$$= 1-x$$

---

11. $3x-2(3x+4)$ **16.** $7a-(a+6)$

12. $5-4(5+x)$ **17.** $10-4(3x+2)$

13. $7c-(c+2)$ **18.** $40-2(1+5w)$

14. $5x-4(2+x)$ **19.** $6y-3(3y+4)$

15. $9-2(4x+1)$ **20.** $8-3(2+5z)$

MULTIPLICATION OF DIRECTED NUMBERS

Consider the expression $6x-(x-3)$.

From $6x$ we have to subtract the number in the bracket which is 3 less than x.

If we start by writing $6x-x$ we have subtracted 3 too many.

To put it right we must add on 3.

Therefore $6x-(x-3) = 6x-x+3$

Similarly $8x-3(x-2)$ means

"from $8x$ subtract three times the number that is 2 less than x".

If we write $8x-3x$ we have subtracted too much, by an amount equal to 3 twos (i.e. 6). So we must add 6 on again giving

$$8x-3(x-2) = 8x-3x+6$$

From this, and from the previous exercise, we have

a)
$$(+3) \times (+2) = +6$$

This is just the multiplication of positive numbers,
i.e. $(+3) \times (+2) = 3 \times 2 = 6$

b)
$$(-3) \times (+2) = -6$$

Here we could write $(-3) \times (+2) = -3(2)$.
This is equivalent to subtracting 3 twos, i.e. subtracting 6.

c)
$$(+4) \times (-3) = -12$$

This means four lots of -3,
i.e. $(-3) + (-3) + (-3) + (-3) = -12$

d)
$$(-2) \times (-3) = +6$$

This can be thought of as taking away two lots of -3,
i.e. $-2(-3) = -(-6)$
We have already seen that taking away a negative number is
equivalent to adding a positive number, so $(-2) \times (-3) = +6$.

EXERCISE 21c

Calculate   a) $(+2) \times (+4)$ b) 2×4.

a) $(+2) \times (+4) = 8$

b) $2 \times 4 = 8$

Calculate   a) $(-3) \times (+4)$ b) -3×4.

a) $(-3) \times (+4) = -12$

b) $-3 \times 4 = -12$

Calculate   a) $(+4) \times (-3)$ b) $4 \times (-3)$.

a) $(+4) \times (-3) = -12$

b) $4 \times (-3) = -12$

> Calculate   a) $(-5)\times(-2)$ b) $-5(-2)$.
>
> a) $(-5)\times(-2) = 10$
>
> b) $-5(-2) = 10$

Calculate:

1.	$(-3)\times(+5)$	**5.**	$(+6)\times(-7)$	**9.**	$(+5)\times(-1)$
2.	$(+4)\times(-2)$	**6.**	$(-4)\times(-3)$	**10.**	$(-6)\times(-3)$
3.	$(-7)\times(-2)$	**7.**	$(-6)\times(+3)$	**11.**	$(-3)\times(-9)$
4.	$(+4)\times(+1)$	**8.**	$(-8)\times(-2)$	**12.**	$(-2)\times(+8)$

13.	$7\times(-5)$	**17.**	$-6(4)$	**21.**	$3(-2)$
14.	$-6(-4)$	**18.**	$-2(-4)$	**22.**	5×3
15.	-3×5	**19.**	$-(-3)$	**23.**	$6\times(-3)$
16.	$5\times(-9)$	**20.**	$4\times(-2)$	**24.**	$-5(-4)$

25.	$6\times(-4)$	**27.**	$(+5)\times(+9)$	**29.**	$7(-4)$
26.	$-3(+8)$	**28.**	-4×5	**30.**	$(-4)\times(-9)$

EXERCISE 21d Multiply out the following brackets:

> $-4(3x-4)$
>
> $$-4(3x-4) = -12x+16$$

> $-(x+2)$
>
> $(-(x+2)$ means $-1(x+2))$
>
> $$-(x+2) = -x-2$$

1.	$-6(x-5)$	**6.**	$-7(x+4)$
2.	$-5(3c+3)$	**7.**	$-3(2d-2)$
3.	$-2(5e-3)$	**8.**	$-2(4+2x)$
4.	$-(3x-4)$	**9.**	$-7(2-3x)$
5.	$-8(2-5x)$	**10.**	$-(4-5x)$

11. $4(3x+9)$	**16.** $-(3x+2)$
12. $5(2+3x)$	**17.** $8(2-3x)$
13. $3(2x-6)$	**18.** $-3(2y-4x)$
14. $-7(2+x)$	**19.** $5(4x-1)$
15. $-2(3x-1)$	**20.** $-5(1-4x)$
21. $6(4+5x)$	**26.** $2(3x+2y+1)$
22. $-6(4+5x)$	**27.** $-5(5+2x)$
23. $6(4-5x)$	**28.** $4(x-y)$
24. $-6(4-5x)$	**29.** $-(4c-5)$
25. $-(5a+5b)$	**30.** $9(2x-1)$

EXERCISE 21e Simplify the following expressions:

1. $5x+4(5x+3)$	**6.** $9-2(4g-2)$
2. $42-3(2c+5)$	**7.** $4-(6-x)$
3. $2m+4(3m-5)$	**8.** $10f+3(4-2f)$
4. $7-2(3x+2)$	**9.** $7-2(5-2s)$
5. $x+(5x-4)$	**10.** $7x+3(4x-1)$
11. $7(3x+1)-2(2x+4)$	**16.** $6x+2(3x-7)$
12. $5(2x-3)-(x+3)$	**17.** $20x-4(3+4x)$
13. $2(4x+3)+(x-5)$	**18.** $4(x+1)+5(x+3)$
14. $7(3-x)-(6-2x)$	**19.** $3(2x+3)-5(x+6)$
15. $5+3(4x+1)$	**20.** $5(6x-3)+(x+4)$
21. $4(x-1)+5(2x+3)$	**26.** $3x+2(4x+2)+3$
22. $4(x-1)-5(2x+3)$	**27.** $5-4(2x+3)-7x$
23. $4(x-1)+5(2x-3)$	**28.** $3(x+6)-(x-3)$
24. $4(x-1)-5(2x-3)$	**29.** $3(x+6)-(x+3)$
25. $8(2x-1)-(x+1)$	**30.** $7x+8x-2(5x+1)$

EQUATIONS CONTAINING BRACKETS

If we wish to solve equations containing brackets we first multiply out the brackets and then collect like terms.

EXERCISE 21f Solve the following equations:

$4+2(x+1)=22$

$$4+2(x+1) = 22$$
$$4+2x+2 = 22$$
$$2x+6 = 22$$

Take 6 from both sides $\qquad 2x = 16$

Divide both sides by 2 $\qquad x = 8$

Check: If $x = 8$, left-hand side $= 4+2(8+1)$
$$= 4+2 \times 9$$
$$= 22$$

Right-hand side $= 22$, so $x = 8$ is the solution.

1. $6+3(x+4) = 24$

2. $3x+2 = 2(2x+1)$

3. $5x+3(x+1) = 14$

4. $5(x+1) = 20$

5. $2(x+5) = 6(x+1)$

6. $28 = 4(3x+1)$

7. $4+2(x-1) = 12$

8. $7x+(x-2) = 22$

9. $1-4(x+4) = x$

10. $8x-3(2x+1) = 7$

11. $16-4(x+3) = 2x$

12. $5x-2(3x+1) = -6$

13. $4x-2 = 1-(2x+3)$

14. $4 = 5x-2(x+4)$

15. $9x-7(x-1) = 0$

16. $16-2(2x-3) = 7x$

17. $3x-2 = 5-(x-1)$

18. $7x+x = 4x-(x-1)$

19. $3-6(2x-3) = 33$

20. $6x = 2x-(x-4)$

21. $3(x+2)+4(2x+1) = 6x+20$

22. $9(2x-1)+2(3x+4) = 20x+3$

23. $3(x+2)+4(2x-1) = 5(x-2)$

24. $2(2x+1)+4 = 6(3x-6)$

25. $6x+4+5(x+6) = 12$

26. $3-4(2x+3) = -25$

27. $15+5(x-7) = x$

28. $6x-2-3(x-4) = 13$

29. $6(x-2)-(2x-1) = 2$

30. $4(2x-5)+6 = 2$

PROBLEMS TO BE SOLVED BY FORMING EQUATIONS

EXERCISE 21g Solve the following problems by forming an equation in each case. Explain, either in words or on a diagram, what your letter stands for and always end by answering the question asked.

The width of a rectangle is x cm. Its length is 4 cm more than its width. The perimeter is 48 cm.
What is the width?

The width is x cm so the length is $(x+4)$ cm.

$$x+(x+4)+x+(x+4) = 48$$

$$4x+8 = 48$$

Take 8 from each side $4x = 40$

Divide each side by 4 $\quad x = 10$

Therefore the width is 10 cm.

(x+4)cm ... x cm ... x cm ... (x+4) cm

A choc-ice costs x pence and a cone costs 3 pence less. One choc-ice and two cones together cost 54 pence.
How much is a choc-ice?

A choc-ice costs x pence and a cone costs $(x-3)$ pence.

$$x+2(x-3) = 54$$

$$x+2x-6 = 54$$

$$3x-6 = 54$$

Add 6 to each side $\quad 3x = 60$

Divide each side by 3 $\quad x = 20$

Therefore a choc-ice costs 20 pence.

---

1. I think of a number, double it and add 14. The result is 36. What is the number?

2. I think of a number and add 6. The result is equal to twice the first number. What is the first number?

3. In triangle ABC, AB = AC.
The perimeter is 24 cm.
Find AB.

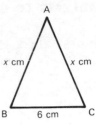

4. I think of a number, take away 7 and multiply the result by 3, giving 15. What is the number?

5. A bun costs x pence and a cake costs 3 pence more than a bun. Four cakes and three buns together cost 40 pence. How much does one bun cost?

6. A bus started from the terminus with x passengers. At the first stop another x passengers got on and 3 got off. At the next stop, 8 passengers got on. There were then 37 passengers. How many passengers were there on the bus to start with?

7. Buns cost x pence each and a cake costs twice as much as a bun. I buy two buns and three cakes and pay 40 pence altogether. How much does one bun cost?

8. I think of a number, add 6, multiply by 2 and the result is 20. What is the number?

9. Jane has x pence and Michael has 6 pence less than Jane. Together they have 30 pence. How much has Jane?

10. The first angle measures $x°$, the second angle (going clockwise) is twice the first, the third is 30° and the fourth is 90°. Find the first angle.

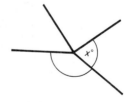

11. 30 sweets are divided amongst Anne, Mary and John. Anne has x sweets, Mary has three times as many as Anne, and John has 6. How many sweets has Anne?

12. In triangle ABC, $\hat{A} = x°$, $\hat{B} = 70°$ and $\hat{C}$ is 20° more than $\hat{A}$. Draw a diagram and find $\hat{A}$.

13. I think of a number, x, and take away 4. Then I start again with x, double it and subtract 8. The two results are the same. What is the number?

INDICES

We have already seen that the shorthand way of writing $2 \times 2 \times 2 \times 2$ is 2^4. In the same way we can write $x \times x \times x \times x$ as x^4. The 4 is called the *index*.

EXERCISE 21h Write the following expressions in index form:

> $y \times y \times y$
>
> $$y \times y \times y = y^3$$

1. $z \times z \times z$ **3.** $b \times b \times b \times b \times b$ **5.** $s \times s \times s$

2. $a \times a$ **4.** $y \times y \times y \times y \times y$ **6.** $z \times z \times z \times z \times z \times z$

Give the meanings of the following expressions:

> z^2
>
> $$z^2 = z \times z$$

7. a^3 **9.** b^2 **11.** x^6

8. x^4 **10.** a^5 **12.** z^4

Simplify the following expressions:

> $2 \times x \times y \times x \times 3$
>
> (Write the numbers first, then the letters in alphabetical order.)
>
> $$2 \times x \times y \times x \times 3 = 2 \times 3 \times x \times x \times y$$
> $$= 6x^2 y$$

13. $2 \times a$ **15.** $3 \times a \times 4$ **17.** $3 \times z \times x \times 5 \times z$

14. $4 \times x \times x$ **16.** $a \times a \times b$ **18.** $5 \times a \times b \times b \times a$

Give the meanings of the following expressions:

> $4ab^2 c$
>
> $$4ab^2 c = 4 \times a \times b \times b \times c$$

19. $3z^2$ **21.** $4zy^2$ **23.** $2x^3$

20. $2abc$ **22.** $6a^2 b$ **24.** $3a^4 b^2$

Simplify the following expressions:

25. $3x \times 2z$ <u>**28.**</u> $3a \times 2a \times a$

26. $x \times 6x^2$ <u>**29.**</u> $a \times b \times c \times 2a$

27. $4a^2 \times 3$ <u>**30.**</u> $4x \times 3y \times 2x$

31. $z \times z \times z \times z$ <u>**37.**</u> $y \times z \times y \times z$

32. $2z \times 3z$ <u>**38.**</u> $2x \times 5z \times y$

33. $4x^2 \times 6$ <u>**39.**</u> $a \times a \times a \times a \times a \times a \times a$

34. $2 \times 4 \times x \times 2$ <u>**40.**</u> $4x^2 \times 2x^2$

35. $4s^2 \times s$ <u>**41.**</u> $x \times y \times z \times a$

36. $x^2 \times x^4$ <u>**42.**</u> $s^4 \times s^3$

MULTIPLICATION AND DIVISION OF ALGEBRAIC FRACTIONS

Algebraic fractions can be multiplied and simplified in the same way as arithmetic fractions.

EXERCISE 21i Simplify the following fractions:

a) $\dfrac{24}{5} \times \dfrac{10}{9}$ b) $\dfrac{2z}{3} \times \dfrac{6}{z^2}$

a) $\dfrac{\cancel{24}^{\,8}}{\cancel{5}_{1}} \times \dfrac{\cancel{10}^{\,2}}{\cancel{9}_{3}} = \dfrac{16}{3}$ b) $\dfrac{2z}{3} \times \dfrac{6}{z^2} = \dfrac{2 \times \cancel{z}^{\,1}}{\cancel{3}_{1}} \times \dfrac{\cancel{6}^{\,2}}{\cancel{z}\times z}_{1}$

$= 5\tfrac{1}{3}$ $= \dfrac{4}{z}$

1. $\dfrac{5}{6} \times \dfrac{12}{5}$ **5.** $\dfrac{a}{4} \times \dfrac{6b}{5}$ <u>**9.**</u> $\dfrac{3c}{5} \times \dfrac{c}{6}$

2. $\dfrac{11}{9} \times \dfrac{18}{5}$ **6.** $\dfrac{2c}{5} \times \dfrac{10}{3c}$ <u>**10.**</u> $\dfrac{4z}{3} \times \dfrac{9}{2z}$

3. $\dfrac{2}{3} \times \dfrac{15}{16}$ **7.** $\dfrac{p}{3} \times \dfrac{9}{p}$ <u>**11.**</u> $\dfrac{5x}{2} \times \dfrac{x}{10}$

4. $\dfrac{z}{3} \times \dfrac{z}{2}$ **8.** $\dfrac{y}{6} \times \dfrac{y}{4}$ <u>**12.**</u> $\dfrac{7}{y} \times \dfrac{2y}{14}$

Division of algebraic fractions follows the same rules as for arithmetic fractions.

Simplify the following fractions:

a) $\dfrac{4}{5} \div \dfrac{8}{15}$ b) $\dfrac{4y}{3z} \div \dfrac{16y^2}{9}$

a) $\dfrac{4}{5} \div \dfrac{8}{15} = \dfrac{\cancel{4}^{\,1}}{\cancel{5}_{\,1}} \times \dfrac{\cancel{15}^{\,3}}{\cancel{8}_{\,2}}$ b) $\dfrac{4y}{3z} \div \dfrac{16y^2}{9} = \dfrac{\cancel{4} \times \cancel{y}^{\,1}}{\cancel{3} \times z}_{\,1} \times \dfrac{\cancel{9}^{\,3}}{\cancel{16} \times \cancel{y} \times y}_{\,4\ \ 1}$

$\qquad\qquad = \dfrac{3}{2}$ $= \dfrac{3}{4yz}$

$\qquad\qquad = 1\dfrac{1}{2}$

13. $\dfrac{1}{3} \div \dfrac{5}{6}$ **17.** $\dfrac{2y}{3} \div \dfrac{y}{6}$

14. $\dfrac{7}{9} \div \dfrac{2}{3}$ **18.** $\dfrac{6}{5y} \div \dfrac{3}{y}$

15. $\dfrac{3}{4} \div \dfrac{7}{12}$ **19.** $\dfrac{4c}{3} \div \dfrac{8y}{9}$

16. $\dfrac{z}{2} \div \dfrac{z}{4}$ **20.** $\dfrac{6z}{25} \div \dfrac{4z^2}{5}$

21. $\dfrac{r}{4} \times \dfrac{r}{6}$ **27.** $\dfrac{16a}{9} \div \dfrac{4ab}{15}$

22. $\dfrac{4y}{3} \times \dfrac{15z}{8y}$ **28.** $\dfrac{yz}{xy} \times \dfrac{vx}{zv}$

23. $\dfrac{3b}{7} \div \dfrac{9ab}{14}$ **29.** $\dfrac{3a^2}{4y} \times \dfrac{2y^2}{6a}$

24. $\dfrac{ab}{bc} \times \dfrac{c}{a}$ **30.** $\dfrac{3}{x} \div \dfrac{6}{y}$

25. $\dfrac{4}{x} \div \dfrac{16}{x^2}$ **31.** $\dfrac{10a}{7} \times \dfrac{14}{5ab}$

26. $\dfrac{3s}{2} \div \dfrac{6s}{7}$ **32.** $\dfrac{3x}{2} \div \dfrac{9y}{4}$

MIXED EXERCISES

EXERCISE 21j **1.** Solve the equation $2x-3 = 7$.

2. Simplify $4(x-3)+1$.

3. I think of a number, double it, add 6 and the result is 32. Find the number.

4. Solve the equation $3x+1 = 2x-3$.

5. What is the meaning of $4a^2$?

6. Solve the equation $4x-5 = 3-2x$.

7. Simplify $3x+2-x-3$.

8. Solve the equation $5-2x = 5$.

EXERCISE 21k **1.** Solve the equation $5x+6 = 3-x$.

2. Simplify $4x-3(2x-5)$.

3. I think of a number, double it, subtract 10 and my answer is 2 more than the number I first thought of. What was the number?

4. Simplify $3a \times 5b \times 4c$.

5. Solve the equation $4(x+3) = 3(2x-4)$.

6. Simplify $\dfrac{2x}{3y} \times \dfrac{6y}{4x}$.

7. Simplify $2x+6y-4x+7x$.

8. Solve the equation $2x-(x+3) = 0$.

EXERCISE 21l **1.** Solve the equation $4-2x = 6-3x$.

2. Simplify $a^3 \times a^3$.

3. Andrew has 6 sweets, Mary has x sweets and Jim has twice as many as Andrew. Together they have four times as many as Mary has. Form an equation and find how many sweets Mary has.

4. Solve the equation $3x-4+8+x = 2x+3-5$.

5. Simplify $3-(x-1)$.

6. Simplify $\dfrac{6x}{5y} \times \dfrac{2y}{3x}$.

7. Simplify $3(x-1)+4(2x+3)-5(x+1)$.

8. Simplify $2(x+3)-4(x-1)$.

EXERCISE 21m **1.** Solve the equation $3x+7 = x+1$.

2. Simplify $\dfrac{4x}{11} \div \dfrac{5x}{22}$.

3. Solve the equation $4-2x = 6-4x$.

4. When shopping, Mrs Jones spent £x in the first shop, the same amount in the second shop, £2 in the third and £8 in the last. The total amount she spent was £18. Form an equation. How much did she spend in the first shop?

5. What is the meaning of x^5?

6. Simplify $x-1-5+4x-x$.

7. Simplify $4-(x-1)$.

8. There is no solution of the equation $4x+3+2x = 6x$. Find the reason by trying to solve the equation.

22 STATISTICS

FREQUENCY TABLES

The branch of mathematics called statistics is used for dealing with large collections of information in the form of numbers. The number of items of information can run into thousands as, for instance, when the incomes of everyone in Britain are being considered, but to learn the methods we start with smaller collections.

If we collect the heights in centimetres of 72 children in the first year we are faced with a disorganised set of numbers:

147	146	151	137	149	159	142	150	151
138	139	155	151	152	145	139	135	153
139	151	145	162	152	138	142	140	155
146	165	155	149	162	145	152	148	152
132	152	142	152	152	143	145	157	152
148	145	154	145	149	155	137	144	140
139	145	151	152	152	140	160	155	151
136	151	149	151	156	142	134	156	156

To make sense of these numbers we must put them in order. One way of doing this is to form a *frequency table*. We do not always wish to write down every number so we group them, as in the next table shown. Work down the columns, making a tally mark, |, in the tally column opposite the appropriate group. (Do *not* go through the columns looking for numbers that fit into the first group and the second group and so on.) Count up the tally marks and write the total in the frequency column. Check by adding up the numbers in the frequency column.

(Arrange the tally marks in fives either by leaving a gap between blocks as is done in this table or by crossing four tally marks with the fifth thus: ⊬⊬⊦ ⊬⊬⊦ ||.)

Height in cm (correct to nearest cm)	Tally	Frequency
131–135	|||	3
136–140	||||| ||||| ||	12
141–145	||||| ||||| |||	13
146–150	||||| |||||	10
151–155	||||| ||||| ||||| ||||| |||||	25
156–160	|||||	5
161–165	||||	4
	Total	72

We can see now that there are a few children with small or large heights and that the greatest number have heights in the 151–155 cm range.

306

EXERCISE 22a Draw up tables like the ones on the previous page, using the groups suggested.

1. The following numbers are the heights in centimetres of the same children as those on page 306, when they had reached the third year:

154	166	153	166	149	154	153	160
165	164	156	166	156	166	161	155
164	164	156	159	161	150	163	
163	154	157	159	150	146	157	
168	167	154	166	150	157	154	
162	164	152	154	153	163	157	
163	161	168	150	152	163	164	
157	159	160	164	158	158	165	
167	170	156	164	164	152	155	
163	164	157	166	161	148	168	

Use groups 146–150 cm, 151–155 cm, 156–160 cm, 161–165 cm, 166–170 cm.

2. The following numbers are the weights in kilograms (to the nearest kg) of the same 72 children in the third form:

41	50	54	52	65	54	48	50
43	48	58	46	43	50	48	47
44	48	43	44	47	45	57	
54	42	52	49	47	40	53	
41	41	49	44	59	43	35	
51	44	44	49	45	62	46	
51	55	54	54	41	43	70	
40	44	59	45	43	45	37	
51	39	55	53	45	61	44	
57	39	51	44	48	44	51	

Use groups 35–39 kg, 40–44 kg, 45–49 kg, 50–54 kg, 55–59 kg, 60–64 kg, 65–69 kg, 70–74 kg.

3. The following are the marks of 82 pupils in a mathematics examination:

78	41	56	66	76	65	50	37	45
40	87	38	49	82	41	79	66	95
19	38	31	75	54	49	65	53	
69	63	67	91	62	34	79	84	
71	85	42	59	74	56	56	50	
53	68	61	54	25	64	84	80	
48	64	72	53	44	55	35	63	
36	81	70	73	47	63	42	57	
51	63	52	45	38	62	64	47	
62	48	28	60	61	58	57	39	

Use groups 11–20, 21–30, 31–40, and so on.

BAR CHARTS

The information collected can be illustrated in various ways and one of the most common is the bar chart.

Here is a bar chart to show the heights of 72 children in the first year (using the table on page 306)

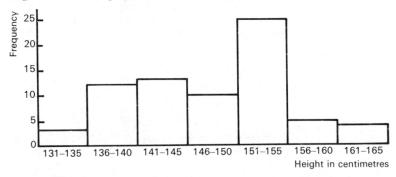

Notice that the groups are arranged along the base line and the frequencies are marked on the vertical axis.

We can see at a glance that the greatest number of children have heights between 151 and 155 cm.

A group of people were asked to select their favourite colour from a card showing 6 colours and the following results were recorded:

Colour	Rose pink	Sky blue	Golden yellow	Violet	Lime green	Tomato red
Number of people (frequency)	6	8	8	2	1	10

The bar chart below shows people's favourite colours:

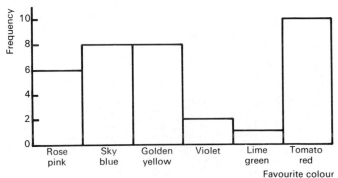

Notice that the bars are all the same width.

EXERCISE 22b In questions 1 to 5 draw bar charts to show the information given in the frequency tables. Mark the frequency on the vertical axis and label the bars below the horizontal axis.

1. Types of vehicles moving along a busy road during one hour:

Vehicle	Cars	Vans	Lorries	Motorcycles	Bicycles
Frequency	62	11	15	10	2

2. Thirty pupils were asked to state their favourite subject chosen from their school timetable:

Subject	English	Mathematics	French	PE	History	Geography
Frequency	5	7	4	3	7	4

3–5. Use the information from the frequency tables in Exercise 22a (page 307), numbers 1 to 3.

6. Population in five towns (to the nearest thousand):

Town	Berkhamsted	Bickley	Brotton	Castle Bromwich	Faversham
Population	15 900	15 800	15 100	16 000	15 000

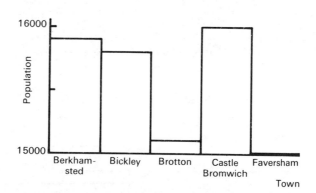

In an attempt to save space, the population scale in the bar chart shown above was started at 15 000 instead of at 0.

a) Redraw the bar chart with the population scale from 0 to 16 000 (suggested scale 1 cm to 1000).

b) Compare the two bar charts. The impression given by one of them is misleading. Why?

Bar charts can be used to represent information other than frequencies and can appear in different forms. The bars are usually vertical but occasionally they are horizontal.

EXERCISE 22c 1.   Average monthly rainfall in Norwich:

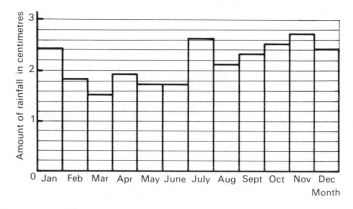

a) Which month is the wettest and which the driest?

b) Does more rain fall in the spring or in the autumn?

c) When would you recommend taking a holiday in Norwich and why?

d) How much rain falls in a year in Norwich?

2.   A rough guide to the distance to keep behind another car on the road:

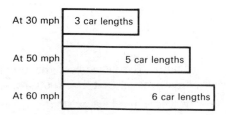

a) What rule has been used to decide on the distances?

b) Why is the guide only "rough"?

3. Bar chart showing the average daily hours of sunshine in Aberdeen and Margate:

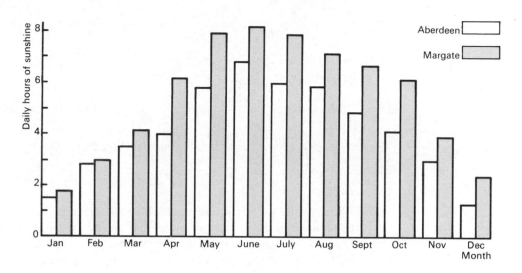

a) Is there more sunshine in Margate or in Aberdeen?

b) Which month is the finest in both towns?

c) Which month has the least sunshine in each town?

4. Cost of fuel in an average home with central heating:

No numbers are given but we can get an idea about the relative costs.

a) Which is the most expensive method overall?

b) Which is the cheapest?

c) Which is the most expensive method of producing hot water?

d) Which is the cheapest method of cooking?

e) Which is the cheapest method of heating?

EXERCISE 22d **1.** a) Draw a bar chart to show the average daily sunshine in Cardiff.

Month	Jan	Feb	Mar	Apr	May	June	July	Aug	Sept	Oct	Nov	Dec
Sunshine in hours	1.7	2.6	4.0	5.5	6.4	7.2	6.2	6.0	4.8	3.4	2.1	1.6

b) In which two months is there least sunshine?

c) In which month is there most sunshine?

2. a) Draw a bar chart to show the average monthly rainfall in Bath.

Month	Jan	Feb	Mar	Apr	May	June	July	Aug	Sept	Oct	Nov	Dec
Rainfall in cm	3.0	2.3	2.0	2.1	2.3	1.8	2.8	2.8	2.6	3.3	3.2	2.9

b) Which is the wettest month and which the driest?

c) Compare this bar chart with the first bar chart on page 310. Does it rain more in Bath than in Norwich?

d) Is it wettest at the same time of year in the two places?

PIE CHARTS

A pie chart is used to represent information when some quantity is shared out and divided into different categories.

Here is a pie chart to show the proportions, within a group, of people with eyes of certain colours.

The size of the "pie slice" represents the size of the group. We can see without looking at the numbers that there are about the same number of people with brown eyes as with grey eyes and that there are about twice as many with grey eyes as with blue. The size of the pie slice is given by the size of the angle at the centre, so to draw a pie chart we need to calculate the sizes of the angles.

The number of people is 60.

As there are 12 blue-eyed people, they form $\frac{12}{60}$ of the whole group and are therefore represented by that fraction of the circle.

Blue: $\frac{12}{60} \times \frac{360°}{1} = 72°$ Grey: $\frac{20}{60} \times \frac{360°}{1} = 120°$

Hazel: $\frac{6}{60} \times \frac{360°}{1} = 36°$ Brown: $\frac{22}{60} \times \frac{360°}{1} = 132°$

Total   360°

Now draw a circle of radius about 5 cm (or whatever is suitable). Draw one radius as shown and complete the diagram using a protractor, turning your page into the easiest position for drawing each new angle.

Label each "slice".

EXERCISE 22e Draw pie charts to represent the following information, first working out the angles.

1. A box of 60 coloured balloons contains the following numbers of balloons of each colour:

Colour	Red	Yellow	Green	Blue	White
Number of balloons	16	22	10	7	5

2. Ninety people were asked how they travelled to work and the following information was recorded:

Transport	Car	Bus	Train	Motorcycle	Bicycle
Number of people	32	38	12	6	2

3. On a cornflakes packet the composition of 120 g of cornflakes is given in grams as follows:

Protein	Fat	Carbohydrate	Other ingredients
101	1	10	8

4. Of 90 cars passing a survey point it was recorded that 21 had two doors, 51 had four doors, 12 had three (two side doors and a hatchback) and 6 had five doors.

5. A large flower arrangement contained 18 dark red roses, 6 pale pink roses, 10 white roses and 11 deep pink roses.

6. Use the information given in Exercise 22b, number 2, page 309.

7. The children in a class were asked what pets they owned and the following information was recorded:

Animal	Dog	Cat	Bird	Small animal	Fish
Frequency	8	10	3	6	3

Sometimes the total number involved is not as convenient as in the previous problems. We may have to find an angle correct to the nearest degree.

If there had been 54 people whose eye colours were recorded we might have had the following information:

Eye colour	Blue	Grey	Hazel	Brown
Frequency	10	19	5	20

Total = 54

Angles Blue: $\dfrac{10}{54} \times \dfrac{360°}{1} = \dfrac{200°}{3}$

$$= 66\tfrac{2}{3}° = 67°\quad\text{(to the nearest degree)}$$

Grey: $\dfrac{19}{54} \times \dfrac{360°}{1} = \dfrac{380°}{3}$

$$= 126\tfrac{2}{3}° = 127°\text{(to the nearest degree)}$$

Hazel: $\dfrac{5}{54} \times \dfrac{360°}{1} = \dfrac{100°}{3}$

$$= 33\tfrac{1}{3}° = 33°\quad\text{(to the nearest degree)}$$

Brown: $\dfrac{20}{54} \times \dfrac{360°}{1} = \dfrac{400°}{3}$

$$= 133\tfrac{1}{3}° = 133°\text{(to the nearest degree)}$$

Total   360°

Draw pie charts to represent the following information, working out the angles first and, where necessary, giving the angles correct to the nearest degree.

8. 300 people were asked whether they lived in a flat, a house, a bedsitter, a bungalow or in some other type of accommodation and the following information was recorded:

Type of accommodation	Flat	House	Bedsitter	Bungalow	Other
Frequency	90	150	33	15	12

9. In a street in which 80 people live the numbers in various age groups are as follows:

Age group (years)	0–15	16–21	22–34	35–49	50–64	65 and over
Number of people	16	3	19	21	12	9

10. Use the information given in Exercise 22b, number 1, page 309.

11. Use the information on people's choice of colour given on page 308.

INTERPRETING PIE CHARTS

EXERCISE 22f 1. This pie chart shows the uses of personal computers in 1981:

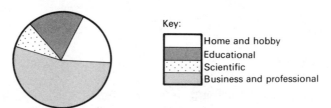

Key:
■ Home and hobby
▨ Educational
░ Scientific
▒ Business and professional

a) For which purpose were computers used most?
b) Estimate the fraction of the total sales used for
  i) scientific purposes
  ii) home and hobbies.

2. The pie chart below shows how fuel is used for different purposes in the average house:

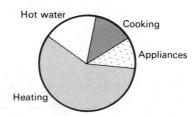

a) For which purpose is most fuel used?
b) How does the amount used for cooking compare with the amount used for hot water?

3. The pie chart shows the age distribution of the population in years in 1964:

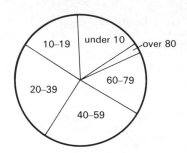

a) Estimate the size of the fraction of the population in the age groups
i) under 10 years ii) 20–39 years.
b) State which groups are of roughly the same size.

PICTOGRAPHS

To attract attention, *pictographs* are often used on posters and in newspapers and magazines. The best pictographs give the numerical information as well; the worst give the wrong impression.

EXERCISE 22g 1. Road deaths in the past 4 years at an accident black spot:

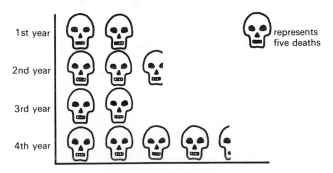

a) Give an estimate of the number of deaths in each year.
b) What message is the poster trying to convey?
c) How effective do you think it is?

2. The most popular subject among first year pupils:

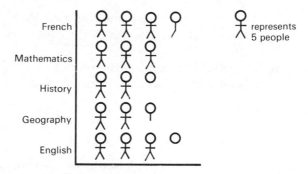

a) Which is the most popular subject?
b) How many pupils chose each subject and how many were asked altogether?
c) Is this a good way of presenting the information?

3. Bar chart in an advertisement showing the consumption of Fizz lemonade:

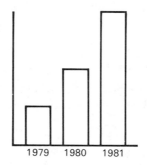

a) What does this show about the consumption of lemonade?

It was decided to change from a bar chart to a pictograph for the next advertisement:

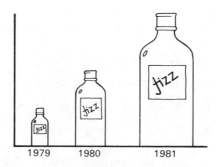

b) This looks impressive but it could be misleading. Why?

DRAWING PICTOGRAPHS

Make sure when using drawings that each takes up the same amount of space and is simple and clear.

EXERCISE 22h **1.** Eighty-five people were asked how they travelled to work and the following information was recorded:

Transport	Car	Bus	Train	Bicycle
Number of people	30	40	10	5

Draw a pictograph using one drawing to 5 people.

2. 30 pupils in a class were asked what they were writing with. The following information was recorded:

Writing implement	Black pen	Blue pen	Pencil
Frequency	12	9	9

3. Some children were asked what pets they owned:

Pet	Dog	Cat	Bird	Small animal	Fish
Frequency	9	7	6	10	2

Use one drawing to one pet. Make the symbols simple.
The symbol for fish could be

PROJECTS

EXERCISE 22i Collect the information; where it is necessary, decide on the groups it can be divided into and record the information in a frequency table as on page 306. Decide whether a bar chart, a pie chart or a pictograph would be most suitable for presenting the information.

Suggestions for Class Projects

1. Heights of children in the class.

2. Weights of children in the class.

3. Handspan. Stretch your hand out as wide as it will go on a piece of paper and mark the positions of the end of the thumb and of the little finger. Measure the distance between these points to the nearest centimetre.

4. Times of journeys to school in minutes.

5. Times of arrival at school.

6. Types of outer garments worn, e.g. blouse or shirt, cardigan, pullover, blazer or jacket.

7. Pets owned.

8. Pets you would *like* to own, but decide on the categories first before collecting the information.

9. Birthday months.

10. Number of houses in the street where a pupil lives. Decide what to record if houses are isolated or the pupil lives in a block of flats.

11. Colours of cars seen passing during, say, 20 minutes.

12. Number of people in cars travelling at a given time of day, say on the way to school.

Suggestions for Individual Projects

13. Toss one dice 120 times and record the scores.

14. Toss two dice 120 times and record the combined score each time.

15. Choose a page of a book of plain text and record the occurrence of the different letters of the alphabet.

16. Choose a page of text in a different language and repeat number 15. Compare the two sets of results.

17. Choose pages of text from a book and record the lengths of, say, 60 sentences.

18. Choose a page of text and record the number of letters used in each word. Decide beforehand what to do about words with hyphens.

19. Count the prickles on holly leaves. Stick to the same bush as there are a number of different varieties of holly.